Biology of AIDS

Second Edition

Wilmore Webley
University of Massachusetts

KENDALL/HUNT PUBLISHING COMPANY
4050 Westmark Drive   Dubuque, Iowa 52002

I dedicate this book to:
All of the past and current students from my Biology of Cancer and AIDS class at the
University of Massachusetts Amherst.

And

My beautiful wife, Ellen Webley, whose constant support has been unwavering.

Contents

Chapter 1: HIV/AIDS: The Early Years 1
    Discovery and Speculations 1
    Current State of the AIDS Epidemic 3

Chapter 2: Understanding HIV and AIDS 7
    What Is AIDS? 7
    What Are Viruses? 7
    The Nature of HIV 9
    HIV Infection and Replication .11

Chapter 3: The Origins of AIDS 15
    HIV/AIDS Dissidents 15
    Evidence that HIV Causes AIDS 17
    How Did HIV First Infect Humans? 19
    SIV and the Origin of AIDS 21

Chapter 4: HIV and the Immune System 23
    Innate and Adaptive Immune Responses 23
    Central Role of CD4 Cells 24
    How Does HIV Attack the Immune System? 26
    Cellular Receptors and Entry of HIV 27

Chapter 5: Progression of HIV Infection to AIDS 29
    Stages of HIV Infection 30
    When Is a Person Diagnosed with AIDS? 31
    AIDS Opportunistic Infections .33
    Common Opportunistic AIDS Infections 33
    AIDS Dementia Complex 37

Chapter 6: Transmission of HIV 39
    Major Routes of HIV Transmission 39
    High-Risk Behaviors 41
    HIV Transmission in MSM 42
    Bug Chasers and HIV Infection 43
    Women and HIV Transmission Risk .44
    Vertical Transmission 45
    Other Possible Modes of HIV Transmission 45
    HIV Infection in Health Care Workers 46
    You Cannot Get HIV From That! 46

Chapter 7: Prevalence of HIV/AIDS 49
    HIV/AIDS in Sub-Saharan Africa 50
    AIDS in Eastern Europe and Central Asia 52
    HIV/AIDS in Oceania 53
    HIV/AIDS in Asia 54
    HIV/AIDS in the Caribbean 55
    HIV Epidemic in Latin America 56
    HIV/AIDS in North America and Western Europe 56
    HIV/AIDS in African Americans 57

Chapter 8: Testing and Diagnosis of HIV/AIDS 61
    HIV Testing for Pregnant Women 62
    Screening and Confirmatory Tests for HIV 62
    Interpretation of Western Blot Data 65
    Nucleic Acid-Based HIV Testing 67
    Rapid HIV Tests 68
    Tests for HIV Progression 70

Chapter 9: HIV/AIDS Treatment 73
    Nucleoside and Non-Nucleoside Inhibitors 73
    Entry Inhibitors 74
    Integrase Inhibitors 75
    HAART 75
    Atripla and AIDS Treatment 77

Chapter 10: HIV Vaccine: A Futile Search? 79
    How Do Vaccines Work? 79
    The Challenges of Creating an HIV Vaccine 81
    Preventative vs. Therapeutic Vaccines 82

Appendix A: Current Selected Issues and Future Outlook 85
    HIV/AIDS and Drug Abuse 85
    Racial and Ethnic Disparities in Infection Rate 86
    Routine HIV Testing as Part of Regular Medical Care 87

Appendix B: Selected HIV/AIDS Online Resources 89

Endnotes 91

CHAPTER 1

HIV/AIDS: The Early Years

Discovery and Speculations

On June 5, 1981, the Centers for Disease Control and Prevention (CDC) in their publication, *Morbidity and Mortality Weekly Report (MMWR)*, described the unusual case studies of five young men[1]. The report covered the period from October 1980 to May 1981 and described the five young men, all active homosexuals, who were treated for biopsy-confirmed *Pneumocystis carinii pneumonia* (PCP) at three different hospitals in Los Angeles, California. Two of the patients died. All five patients had laboratory-confirmed previous or current cytomegalovirus (CMV) infection and mucosal Candidiasis (yeast) infection. The authors of the *MMWR* report, led by Dr. Micheal Gottleib at the University of California at Los Angeles, indicated that all of these patients exhibited signs of severe immunodeficiency not normally seen in this age group. The *New York Times* published an article on July 3, 1981, entitled "Rare Cancer Seen in 41 Homosexuals" describing the unusual occurrence of a rare skin cancer, Kaposi's sarcoma, seen in homosexual men. A day later, on July 4, the CDC reported that during the preceding 30 months, 26 cases of Kaposi's sarcoma had been reported among gay males, and that eight had died, all within 24 months of diagnosis. These reported infections, both in the *MMWR* and the *New York Times*, were later determined to be some of the first cases of infection with the <u>h</u>uman <u>i</u>mmunodeficiency <u>v</u>irus (HIV), which eventually leads to a state of dysfunction in the body's immune functions called <u>a</u>cquired <u>i</u>mmuno<u>d</u>eficiency <u>s</u>yndrome (AIDS); thus began what would become the most deadly pandemic of the 21st century.

Since 1981 more than 60 million people worldwide have been infected with the virus including the approximately 33.2 million people currently living with the disease. While 1981 is generally referred to as the beginning of the HIV/AIDS epidemic, scientists now believe that HIV was present years before the first case was brought to public attention. The U.S. president at that time, Ronald Reagan, refused to talk about the growing problem of AIDS. Nonetheless, in 1982 the first congressional hearings were held on HIV/AIDS. For a while the disease term "GRID" or "gay-related immune deficiency" was used by the media and health care professionals, as it was mistakenly thought that there was an inherent link between homosexuality and the syndrome, and only homosexuals were susceptible. The CDC convened a meeting of scientists, blood industry executives, gay activists, hemophiliacs, and others to develop guidelines for screening the blood supply. Although they decided then to adopt a "wait-and-see" attitude, they formally established the term *acquired immune deficiency syndrome (AIDS)* in 1982, referring at that time to four "identified risk factors" of male homosexuality, intravenous drug abuse, Haitian origin, and hemophilia A. Doctors agreed that the term AIDS was appropriate since people acquired the condition rather than inherited it; because it resulted in immune system deficiency; and because it was a syndrome, characterized by a number of manifestations, rather than a single disease. On April 22, 1984, Dr. James Mason of the CDC was reported as saying, "I believe we have the cause of AIDS." He was referring to a virus called lymphadenopathy virus (LAV) isolated in 1983 by Luc Montagnier of the Pasteur Institute in France. A day later, United States Health and Human Services Secretary Margaret Heckler announced that Dr. Robert Gallo (See Figure 1-1) of the National Cancer Institute had isolated the virus that caused AIDS, a virus Gallo named human T-cell leukemia virus type 3 (HTLV-III).

Heckler stated that there would soon be a commercially available test able to detect the virus and that a vaccine to combat the virus would be available for testing within two years. Heckler concluded *"yet another terrible disease is about to yield to patience, persistence and outright genius."* The virus was later named the human immunodeficiency virus (HIV). Still very little was known about modes of transmission and public anxiety continued to grow. In the same year, the CDC issued a statement suggesting that abstention from intravenous drug use and reduction of needle-sharing should be effective in preventing transmission of the virus. This proved to be correct as time passed. It was not until September 17, 1985, that then-president Ronald Reagan mentioned the word "AIDS" in public for the first time in response to a reporter's query. In this same year, the Food and Drug Administration (FDA) approved the first HIV antibody test and routine tests on blood products in the United States and Japan com-

Figure 1-1 Robert Gallo, co-discoverer of the HIV virus, in the early 1980s among (from left to right) Sandra Eva, Sandra Colombini, and Ersell Richardson.
Source: NIH

menced. It was also in 1985 that the first International Conference on AIDS was held in Atlanta, Georgia, and thus commenced the intense collaborations between and among several agencies of the U.S. government (*http://www.avert.org/his8186.htm*).

Current State of the AIDS Epidemic

Today, 27 years after the first report of a handful of cases of a nameless, deadly disease among male homosexuals in Los Angeles and New York, there are over 1 million persons living with HIV in the United States. About one-fourth of those with HIV have not yet been diagnosed and are unaware that they are infected. Since those very early beginnings we have learned much about HIV, the infectious cause of AIDS, as well as about AIDS, the disease. Consequently, HIV has become the most widely studied virus (See Figure 1-2), and as a direct result of the extensive genetic and proteomic analyses performed on the virus we currently have over 27 FDA-approved drugs that are used to treat HIV infections in the United States.

Unlike the disorganized panic that ensued in the early 1980s, current investigators understand a great deal about the spread of the virus in our population. Most notably, it is now clear that HIV transmission requires close contact with an infected person and that infection occurs most frequently through the exchange of body fluids. This has led many to speculate that AIDS can now be viewed as a chronic disease, just like diabetes, that is quite manageable. While this might be partially true in some developed parts of the world where antiretroviral drugs are readily available and patients can afford them, a quick glance at the worldwide incidence, prevalence, and mortality statistics indicate that we are still faced with a monumental global health crisis. According to the latest data from the Joint United Nations Program on HIV/AIDS (UNAIDS) and the World Health Organization (WHO), as of the end of 2007, 33.2 million people were estimated to be living with HIV/AIDS worldwide. Although the number of people living with the disease continues to increase, the current estimates indicate that the global HIV/AIDS prevalence rate (the percent of people living with the disease) has leveled off. The currently reported estimate is a reduction of 16% compared with the estimate published in 2006 (39.5 million). An estimated 2.5 million people will become infected with HIV in 2007 and approximately 2 million people died from AIDS-related disease last year. Just about half of the adults living with HIV/AIDS worldwide are women, while young people under the age of 25 are estimated to account for half of all new HIV infections

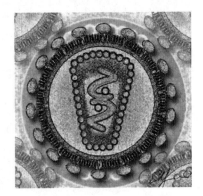

Figure 1-2 Stylized rendering of a cross section of the AIDS virus. *Source: NIH.*

worldwide. It is estimated that every day over 6,800 persons become infected with HIV and over 5,700 die from AIDS. Notably, nearly 70% of the people with AIDS live in sub-Saharan Africa, although this region has only 10% of the world's population. This region is also home to almost 95% of the world's AIDS orphans. For most of the AIDS patients in sub-Saharan Africa and Southeast Asia, there is no adequate health care system and antiviral drugs are either not available or are too expensive for the average person. AIDS has killed more than 25 million people since it was first recognized in 1981, making it one of the most destructive epidemics in recorded history (Statistics from UNAIDS). HIV/AIDS is the leading cause of death worldwide for people aged 15–59, and half of all new infections occur in people under the age of 25 (*http://www.pbs.org/wgbh/ pages/ frontline/aids/etc/synopsis.html*).

HIV is one of the most deadly viruses humankind has ever faced. But, it has one weakness: Infection by the virus is absolutely preventable by relatively basic lifestyle measures. These include using condoms and clean needles, proper testing and protection of the blood supply, and testing and administering of antiretroviral medicine to expectant mothers. After more than a quarter of a century of research and drug development, the inescapable truth is that the only way to win the battle with HIV and AIDS is to reduce the number of people becoming infected in the first place. Currently, prevention is still the only "cure" available for HIV/AIDS. Many experts and government officials therefore agree that a comprehensive approach must be adopted to prevent further spread of HIV and AIDS. HIV prevention strategies must therefore include close monitoring of the epidemic to target prevention and effective patient care activities, researching the effectiveness of prevention methods, disseminating proven effective interventions, funding the implementation and evaluation of prevention efforts in high-risk communities, encouraging early diagnosis of HIV infection, and fostering meaningful collaborations between prevention and treatment programs. Although 95% of all infections take place in the developing world with less organized and funded health care systems, HIV today is a threat to men, women, and children on all continents of the world. For this reason, World AIDS Day (See Figure 1-3 for World AIDS Day Logo), which started on the first of December 1988, is not just a fundraising event, but a deliberate attempt to increase awareness, fight prejudice, and improve HIV/AIDS education. This day is necessary to remind people that HIV has not gone away, and that there is much to be done yet.

Figure 1-3 World AIDS Day is December 1st of each year. Image © kmolnar. Used under license from Shutterstock, Inc.

CHAPTER 2

Understanding HIV and AIDS

What Is AIDS?

As previously stated, AIDS is the acronym for acquired immunodeficiency syndrome. *Acquired*, which means you catch it; *immunodeficiency* means that your immune system is weakened and can no longer fight off infectious agents; and *syndrome*, describing a combination of symptoms that collectively point to a particular disorder. AIDS is believed to be caused by the human immunodeficiency virus (HIV). By leading to the destruction and/or functional impairment of cells of the immune system—notably helper T cells—HIV progressively destroys the body's ability to fight common infections and certain cancers. If left untreated, AIDS leads to death. Nobody dies directly from infection with HIV itself, but from the illnesses that can develop due to the destruction of the immune system by the virus. The term *AIDS* therefore refers to an advanced stage of HIV infection, when the immune system has sustained substantial damage, thereby leaving the body open to opportunistic infections. This means, therefore, that not everyone with HIV has AIDS. We now know that it takes approximately 10 years from the time of initial infection with HIV for the onset of the symptoms of AIDS to appear.

What Are Viruses?

When the word *virus* is mentioned today, most people automatically think about their computers and terms like McAfee and Norton antivirus software start popping into our heads. As I am writing this paragraph, McAfee security center just popped up a message alerting me to a possible virus threat on my computer. For those born before the age of computer viruses, they might think of virus pandemics caused by smallpox, rubella, and influenza. In 1884, Charles Chamberland, in the lab of Louis Pasteur, discovered that if you passed liquid containing bacteria through an unglazed porcelain tube, the bacteria would be retained and the solution that passed through (the filtrate) was sterile. This technique became a major form of sterilization until 1892 when Dmitri Iwanowski applied this test to a filtrate of plants suffering from Tobacco Mosaic Disease with surprising results. The filtrate was capable of producing the original disease in new plants! Repeated filtrations produced the same results and nothing could be seen in the filtrates using the most powerful microscopes at that time. They also realized that they could not culture anything from the filtrates using nutrient agar in Petri plates as they were accustomed to doing with filtered bacteria. Iwanowski and his colleagues consequently concluded that they had discovered a new pathogenic life-form, which they called "Filterable Virus." A virus in this context is an ultramicroscopic (very small), biological infectious agent that lacks the means for self-reproduction outside a host cell. Viruses are therefore called obligate intracellular, but unlike other obligate intracellular parasites (such as *Chlamydia*), viruses are not truly alive, although some may disagree. Although many viruses are pathogenic (cause disease), most viruses in nature do not cause disease in humans. There are viruses that infect animals, plants, insects, and even bacteria (bacteriophages). You can go to *http://en.wikipedia.org/wiki/Listofviruses* for a list of known viruses and the diseases they cause.

At the most basic level, a virus consists of a piece of nucleic acid material (DNA or RNA; never both!), wrapped in a thin coat of protein called the viral capsid. The nucleic acid material can either be double- or single-stranded. Many, but not all, viruses also have an outer coat called an envelope derived from the infected cell membrane. They range in size from 20 to 250 nanometers in diameter (one nanometer is one billionth of a meter). Compare this size to a typical bacterial cell (prokaryotic cell), which measures 1 to 10 micrometers, or the average animal cell (eukaryotic cell), which is approximately 10 to 100 micrometers in diameter (1 micrometer = 1,000 nanometers). See Figure 2-1. Outside of a living cell, a virus is dormant, but once inside, it "injects" its nucleic acid material into the host cell's DNA and takes over the resources of the host cell initiating the production of more virus particles. A virus particle first attaches to a receptor protein on the host cell membrane using an attachment protein that is essential to its survival.

If a virus cannot attach to the cell, it cannot enter and therefore cannot cause disease. These proteins on the host cell to which viral particles attach were not created for viral entry and many times serve very important cellular functions. The viruses have just evolved to use already-existing receptors to enter the cell.

Following attachment, the virus fuses with the cell membrane and extrudes its nucleic acid material into the cell where it integrates into the cellular genome. The virus then utilizes cellular transcription, translation, as well as energy machinery to replicate its genome and produce new viral proteins. The newly formed viral components assemble to form new virions or viruses. Some viruses exit the cells by budding through the cell membrane, thereby acquiring a lipid bilayer in the process. Other

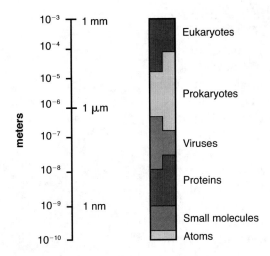

Figure 2-1 The range of sizes shown by viruses, relative to those of other organisms and biomolecules. (Source: created by Tom Vickers (released as public domain))

viruses, however, exit the cell by releasing enzymes that cause cell lysis or rupture and do not acquire a lipid bilayer. The viruses that bud from the cell membrane are called enveloped viruses and those that cause host cell rupture to exit the cell are said to be non-enveloped or naked viruses. A single infected host cell may produce up to 10,000 new virus particles. A virus therefore does not really intend to cause harm; all it wants to do is make more viruses. Unfortunately for us, in its attempt to preserve itself, harm is indeed caused to the individual infected, as well as uninfected host cells, leading to disease pathologies.

The Nature of HIV

HIV hails from the *Lentivirus* family of viruses, all of which are retroviruses and share distinct characteristics, including the ability to systematically attach cells of the immune system. The viruses can be found in a number of different animals, including cats, sheep, horses, and cattle. Interestingly, the simian immunodeficiency virus (SIV) that affects monkeys is a lentivirus. These viruses are known to steadily weaken the body's defense (immune) system until it can no longer fight off infections such as pneumonia, diarrhea, tumors, and other illnesses, all of which can be part of the AIDS (acquired immunodeficiency syndrome) complex.

An HIV particle is approximately 100–150 billionths of a meter in diameter. That is approximately 0.1 microns or 4 millionths of an inch! This means that HIV is one-seventieth of the diameter of the human CD4+ white blood cells that they typically infect (see TEM image in Figure 2-2). These cells are called CD4 because of a protein that is ubiquitously expressed on the cell surface of T-helper cells. HIV particles surround themselves with a viral envelope through which project approximately 72 spikes that are formed from the gp120 and gp41 proteins (Figure 2-3).

The "gp" in gp120 or gp41 refers to the **glycoprotein** nature of the molecules and infers that they are a combination of carbohydrate and protein molecules joined together by chemical linkage. Immediately below this outer viral envelope is the viral matrix, which is made from the protein p17. The viral core (or capsid) is usually bullet-shaped and is made from the p24 protein. Inside the core are three enzymes required for HIV replication called reverse transcriptase, integrase, and protease. Also held within the core is HIV's genetic material. While most organisms, including viruses, store their genetic material on strands of DNA, Retroviruses like HIV are the exception because their genes are composed of RNA (ribonucleic acid). The HIV genome consists of two identical strands of RNA, making the HIV replication process more complicated than that seen in most other viruses.

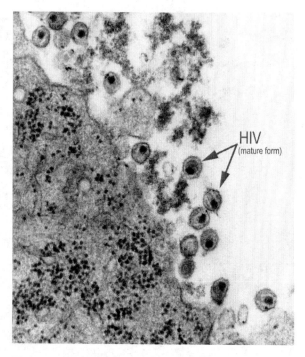

Figure 2-2 This highly magnified transmission electron micrographic (TEM) image reveals the presence of mature forms of the human immunodeficiency virus (HIV) in a tissue sample under investigation. (Source: CDC Public Health Image Library *http://phil.cdc.gov/phil/home.asp.*)

HIV is on the smaller end of the scale where viruses are concerned. Compared to bacteria that generally have over a thousand genes and a human cell containing approximately 30,000 genes, HIV has just nine genes. Three of these HIV genes—*gag, pol,* and *env*—contain the information needed to make structural proteins (i.e., envelope glycoproteins, capsid, and polymerase, respectively) for new virus particles. The remaining six HIV genes, known as *tat, rev, nef, vif, vpr,* and *vpu,* code for proteins that control the ability of HIV to infect a cell, produce new copies of virus, or cause disease.

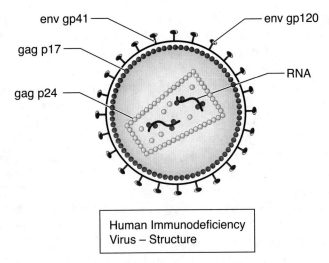

Figure 2-3 This illustration shows the structure of the human immunodeficiency virus (HIV). The outer shell of the virus is known as the viral envelope. Embedded in the viral envelope is a complex protein known as env, which consists of an outer protruding cap glycoprotein (gp) 120, and a stem gp41. Within the viral envelope is an HIV protein called p17 (matrix), and within this is the viral core or capsid, which is made of another viral protein p24 (core antigen). The major elements contained within the viral core are two single strands of HIV RNA, a protein p7 (nucleocapsid), and three enzyme proteins, p51 (reverse transcriptase), p11 (protease) and p32 (integrase). (Source: Courtesy of AVERT.org. *http://www.avert.org/virus.htm.*)

HIV Infection and Replication

Since HIV can only replicate inside human cells, the initial step in its infection involves binding to a susceptible host cell. In fact, for any pathogen to cause disease, it has to first devise a mechanism by which to attach to the host cell. Once attachment is accomplished, the pathogen may gain entry into the cell by various pathways. Although CD4+ T cells appear to be the main targets of HIV, other immune system cells may be infected as well. Cells of the mononuclear phagocyte system, principally blood monocytes and tissue macrophages, T lymphocytes, B lymphocytes, natural killer (NK) lymphocytes, dendritic cells (Langerhans cells of epithelia and follicular dendritic cells in lymph nodes), hematopoietic stem cells, endothelial cells, microglial cells in brain, and gastrointestinal epithelial cells are all targets of HIV infection. Among these, the long-lived monocytes and macrophages have been shown to harbor large quantities of the virus without being killed. These cells therefore function as reservoirs of HIV. CD4+ T cells also serve as important reservoirs of HIV; a small proportion of these cells harbor HIV in a stable, inactive form. Normal immune processes may activate infected cells, resulting in the production of new HIV virions. This attachment event involves the specific interaction between the gp120 spike on HIV (see Figure 2-3) and CD4, an essential protein found on the surface of certain human immune cells.

What typically happens is that a virus particle gains entry to the body through infected body fluids, where it soon bumps into a cell that carries on its surface the CD4 protein. The gp120 spikes on the surface of the virus particle bind to the CD4 molecule allowing the viral envelope to fuse with the host cell membrane. In addition, the viral protein then binds to a second kind of cell-surface molecule known as **chemokine receptors**. The HIV particle must interact with this second molecule on the cell surface to gain entry and commence a productive infection cycle. This second molecule is either **CXCR4** or **CCR5** depending on the cell type. The macrophage-tropic (M-tropic) HIV-1 strains utilize the chemokine receptor CCR5 in combination with CD4 for entering target cells, while the T-tropic HIV-1 uses CXCR4[2]. These molecules are therefore referred to as coreceptors for HIV. In fact, individuals in the population who are missing the gene for CCR5-(Delta32), do not produce a functional protein and exhibit a near-complete protection against HIV-1 infection. The virus then sheds its envelope and the nucleic acid material (RNA) is released into the cell where it is converted by the enzyme **reverse transcriptase (RT)**. The process is called reverse transcription because one typically goes from DNA to RNA and then to protein. However, since HIV is an RNA virus and needs to integrate its genes into the host cell DNA, this enzyme transcribes the single-stranded RNA into double-stranded DNA in the cytoplasm of the cell. The viral DNA then migrates into the nucleus, where it gets integrated into the host DNA by another HIV enzyme called **integrase**. HIV DNA that enters the DNA of the cell is called a **provirus**. It is the job of the provirus to produce new viruses, which will be released from the infected cells. For this to happen, RNA copies must be made that can be read by the host cell's protein-making machinery. These copies are called **messenger RNA (mRNA)**, and production of mRNA is called **transcription**, a process that involves the host cell enzymes. Viral genes in conjunction with the host cell machinery control this process. Viral genomic RNA is also transcribed for later incorporation into the newly assembled virus particle (Figure 2-4).

Following the transcription of viral mRNA in the cell's nucleus, this mRNA is transported to the cytoplasm where protein synthesis will take place. The HIV protein encoded by the _rev_ gene is important here since it allows mRNA encoding the HIV structural proteins to migrate from the nucleus to the cytoplasm. Without the rev protein, structural proteins are not made. In the cytoplasm, the virus co-opts the cell's protein-making machinery to make long chains of viral proteins and enzymes, using HIV mRNA as a template in a process called translation. Newly synthesized HIV core proteins along with enzymes and genomic RNA are assembled inside the cell forming an immature viral particle. This immature virus then buds off from the cell, acquiring

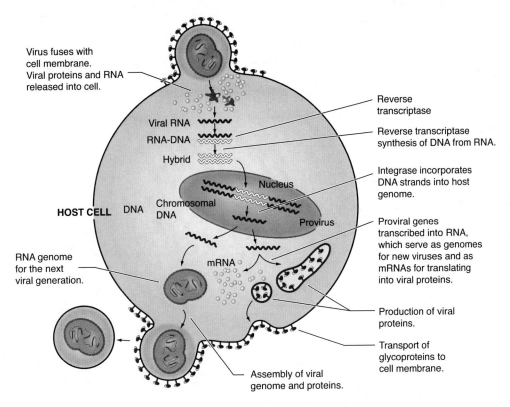

Figure 2-4 Basic schematic diagram of the HIV replication cycle.

an envelope in the process that includes both cellular and viral proteins. During this part of the viral life cycle, the core of the virus is immature and the virus is not yet infectious. The long chains of proteins and enzymes that make up the immature viral core are now cut into smaller pieces by a viral enzyme called protease. This step results in infectious viral particles. One of the reasons HIV is so dangerous is that once it enters the CD4 cell, it hijacks the cell's machinery and turns out over 10 billion to 1 trillion new virus particles per day.

It was recently discovered that viral budding from the host cell is much more complex than previously thought. Binding between the HIV Gag protein and molecules in the cell directs the accumulation of HIV components in special intracellular sacks, called **multivesicular bodies (MVB)**, which normally function to carry proteins out of the cell. In this way, HIV actively hitchhikes out of the cell in the MVB by hijacking normal cell machinery. Discovery of this budding pathway has revealed several potential points for intervening in the viral replication cycle.

CHAPTER 3

The Origins of AIDS

For over 20 years the origin of AIDS has been the subject of fierce debate and the cause of countless arguments and conspiracy theories. The first recognized cases of AIDS in the United States occurred in the early 1980s. These cases plunged the disease in the spotlight when several gay men in San Francisco and New York suddenly began to develop rare opportunistic infections and rare cancers that were uncharacteristically resistant to conventional treatment regimens. Doctors soon discovered a distinctive feature of these cases. More than anything else, the men were lacking a specific type of white blood cell, which is essential to a healthy immune system. Healthy individuals typically have approximately 1,500 "CD4 cells" (also called T-helper cells) in each cubic milliliter of their blood. However, the men with this strange new disease typically had much lower levels, approximately 200. This immune deficiency explained why they were so vulnerable to disease. It soon became clear that all the infected men were suffering from a common syndrome. The disease was formally called AIDS in 1982 and the discovery of HIV, the human immunodeficiency virus that causes AIDS, was made soon after. Some maintain that the virus was developed by the U.S. government to eliminate African Americans and gay people in New York City; others claim that the virus was genetically engineered to wage a biological warfare on Cuba, but then escaped and entered the general population; still others contend that HIV was developed through random mutations arising from secret electromagnetic warfare waged by the Soviet Union against the United States; while many still believe that the virus was spread either by a promiscuous flight attendant or through the oral polio vaccine program. So, what is the truth? Just where did AIDS come from?

HIV/AIDS Dissidents

While most researchers today believe that the evidence supporting the fact that HIV causes AIDS is abundant and conclusive, there are many who disagree with this view. This has sparked many debates and has given rise to several groups of HIV/AIDS dissidents around the world. **AIDS dissident movement,** also referred to as **AIDS**

denialism, is a loosely connected group of individuals who dispute the **scientific consensus** that HIV is the cause of AIDS. Many dissident groups publicly deny the existence of HIV, while others accept that HIV exists, but argue that it is a harmless **passenger virus** and could therefore not be the cause of AIDS. Some dissidents further argue that the consensus that HIV causes AIDS has itself resulted in numerous inaccurate diagnoses leading to psychological terror and toxic treatments (referring here to AZT). They further charge that the commonly held view that HIV causes AIDS has led to a squandering of public funds, as well as an unprecedented deviation from the **scientific method** and standards. To date, the most vocal and probably the most significant scientist to question the HIV/AIDS theory is **Professor Peter Duesberg** (See Figure 3-1), at the University of California at Berkeley.

Dr. Duesberg, a virologist, first published his opinion on this topic in 1988[3]. Even with the large volume of data amassed by HIV/AIDS researchers throughout the 1990s to the present time, Dr. Duesberg remains unconvinced and is probably more defiant. While he concedes that HIV exists, he still maintains that it is harmless and is only present because the patient's immune system is already compromised by other factors[4]. Dr. Duesberg believes that AIDS is caused mainly by drug abuse in developed countries of the world including the use of AZT, an antiviral drug currently used to combat HIV. For AIDS patients living in developing countries, like those on the African continent, he believes that malnutrition is the major factor at work here (*http://www.duesberg.com/*). Dr. Duesberg is not alone; there are many more groups and individuals who believe as he does. Other prominent advocacy groups include the Perth Group led by **Dr. Eleni Papadopulos** (*http://www.theperthgroup.com/*), who do not believe that HIV really exists. The Perth group

Figure 3-1 AIDS dissident Peter Duesberg has been a major force against the HIV/AIDS link. (Source: *http://www.reviewingaids.org/awiki/ index.php/AIDS_dissident*)

maintains that AIDS and all the associated HIV-induced phenomena are in fact caused by changes in cellular redox levels brought about by the oxidative nature of substances and exposures common to all the AIDS risk groups and to the cells used in the so-called "culture" and "isolation" of "HIV." Despite the sometimes confusing message of these dissident groups, there is more than enough evidence to support the fact that the

HIV infection indeed leads to the eventual development of AIDS. Several prominent scientists once associated with AIDS dissident groups have since changed their views and accepted the fact that HIV plays a role in causing AIDS, in response to an accumulation of new data[5]. Probably most noteworthy of this group is **Robert Root-Bernstein,** author of *Rethinking AIDS: The Tragic Cost of Premature Consensus* and formerly a critic of the HIV/AIDS paradigm, has since distanced himself from the AIDS dissident movement, saying, "Both the camp that says HIV is a pussycat and the people who claim AIDS is all HIV are wrong. . . . The denialists make claims that are clearly inconsistent with existing studies. When I check the existing studies, I don't agree with the interpretation of the data, or, worse, I can't find the studies [at all]"[6].

Evidence that HIV Causes AIDS

Among the numerous criteria used over the years to establish a causative link between a putative pathogenic (disease-causing) agent and a disease, the most cited are **Koch's postulates,** a set of rules for the assignment of a microbe as the cause of a disease. These rules were developed in the late 19th century by Robert Koch. Koch was a German physician who became famous for his discovery of the bacilli that cause anthrax, as well as tuberculosis and cholera in the late 1800s. Robert Koch is considered one of the founders of bacteriology and was awarded the Nobel Prize in Physiology and Medicine for his tuberculosis findings in 1905. For more than a century, Koch's postulates have served as the litmus test for determining the cause of any epidemic disease. The basic tenets of the postulates are as follows:

1. The suspected pathogen must always be found in diseased individuals, but absent in a healthy individual.

2. The pathogen must be isolated from the diseased individual and grown in a pure culture.

3. The pathogen from the pure cultures must cause the disease when inoculated into a healthy, susceptible host.

4. The same pathogen must be re-isolated from the host that was inoculated with the pure culture.

In order to address the question of whether HIV really causes AIDS, scientists have applied Koch's postulates as have been done throughout the past two centuries, although we admit that these postulates have inherent shortcomings. With regard to the first postulate, numerous studies from around the world show that HIV has been isolated from virtually every patient with AIDS. In addition, these patients are HIV-seropositive; meaning that they carry antibodies (highly specific immune defense proteins) against HIV, indicating that an HIV infection has taken place. These studies also demonstrated that the greater the HIV **viral load** (number of viral particles in the peripheral blood), the greater the number of patients who develop AIDS within a six-year period (Table 3-1).

With regard to postulate 2, which requires that the pathogen be isolated in pure culture, modern culture techniques have allowed the isolation of HIV in virtually all AIDS patients, as well as in almost all HIV-seropositive individuals (those who have HIV-specific antibodies) with both early- and late-stage disease. In addition, the polymerase chain (PCR) and other molecular techniques that look for viral nucleic acid have enabled researchers to document the presence of HIV genes in virtually all patients with AIDS, as well as in individuals in earlier stages of HIV disease.

Postulate 3 requires the isolated virus to be placed back into an uninfected host and observation of the same disease symptoms as seen in the previous host. This cannot be deliberately done by research scientists because of obvious ethical considerations. However, unfortunate accidental exposure to HIV in lab workers resulted in development of AIDS or severe immunosuppression after accidental exposure to concentrated, cloned HIV in the laboratory when no other risk factors were present.

In all cases, HIV was isolated from the infected individual, sequenced and shown to be the same as the infecting strain of virus. In another tragic incident, transmission of HIV from a Florida dentist to six of his patients has been documented by genetic analyses of viruses isolated from both the dentist and the patients. The dentist and three of the patients developed AIDS and died, and at least one of the other patients has developed AIDS. Five of the patients had no HIV risk factors other than multiple visits to the dentist for invasive procedures[8,9]. The development of AIDS following known HIV infection through seroconversion also has been repeatedly observed in pediatric as well as adult blood transfusion cases. Infections caused by mother-to-child

Table 3-1 The greater the plasma RNA concentration indicating the presence of the virus, the greater the proportion of patients who progressed to AIDS within six years, indicating that there is a direct link between HIV viral load and onset of AIDS Symptoms. Modified from Mallors et al.[7]

Plasma RNA concentration (copies/mL of blood)	Proportion of patients who developed AIDS within 6 years
<500	5.4%
501–3,000	16.6%
3,001–10,000	31.7%
10,0001–30,000	55.2%
>30,000	80.0%

transmission, accidental infection of hemophilia patients, injection-drug use, and sexual transmission in which seroconversion can be documented using serial blood samples, consistently document the fact that HIV isolated from infected individuals can lead to AIDS in previously uninfected individuals[10].

Finally, and maybe most importantly, no other agent or condition, including viral infections, bacterial infections, sexual behavior patterns, and drug abuse patterns predict who develops AIDS besides infection with HIV. Individuals from diverse backgrounds, including heterosexual men and women, homosexual men and women, hemophiliacs, sexual partners of hemophiliacs and transfusion recipients, injection-drug users, the elderly in our society, and infants have all developed AIDS, with the only common denominator being their prior infection with the agent that leads to HIV/AIDS (*http://www.niaid.nih.gov/publications/hivaids/hivaids.htm*). With this accumulated evidence clearly showing that AIDS does not occur in the absence of HIV, that HIV infection is the only universal factor that predicts who will develop AIDS, coupled with surveillance statistics and the tremendous success of modern antiretroviral therapy, it is clear that HIV is the etiologic agent of AIDS.

How Did HIV First Infect Humans?

Two distinct types of HIV can be identified genetically and antigenically. HIV-I is the cause of the current worldwide pandemic and was first observed in 1984, while HIV-2 is found primarily in West Africa but rarely elsewhere. Both HIV-1 and HIV-2 are thought to have arisen from simian immunodeficiency virus (SIV). HIV-2 is more closely related to the SIV found in West Africa. Both HIV-1 and HIV-2 have the same modes of transmission and are associated with similar opportunistic infections and AIDS.

In persons infected with HIV-2, immunodeficiency seems to develop more slowly and tends to be milder. HIV-2 infected individuals are also less infectious early in the course of infection. HIV-1 and HIV-2 also differ in geographic patterns of the disease; the United States has few reported HIV-2 cases, but otherwise clinically the diseases are very similar. There are three subgroups of HIV-1—M (main or major), N (new), and O (outlier). Type O HIV-1 is mostly found in Cameroon and Gabon while the rare N subgroup is also found in Cameroon.

Scientists believe that SIV might have infected humans on separate occasions to give rise to the three subgroups. Currently, there are approximately ten different HIV-1 subtypes within group M (Figure 3-2). Subtype B is the major subtype found to infect populations in North America, Latin America and the Caribbean, Europe, Japan, and Australia. Almost all of the M subtypes are found in sub-Saharan Africa. Subtypes A and D are found at the highest rates in central and eastern Africa and C in southern Africa (Figure 3-3). Type C is the predominant form found in India and also causes the most infections worldwide.

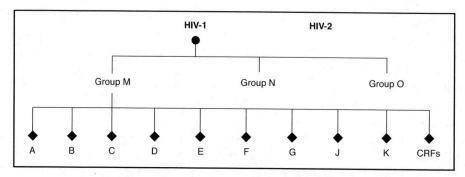

Figure 3-2 Diagram illustrating HIV classification levels. HIV-1 is divided into its representative groups and group M is divided into subtypes and circulating recombinant forms (CRFs).

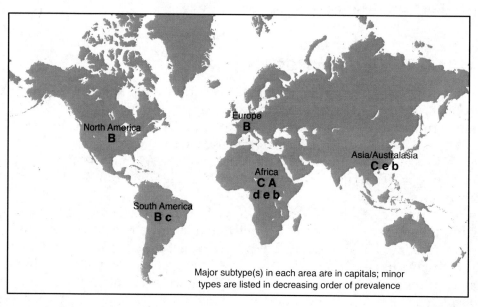

Figure 3-3 Worldwide distribution of HIV subtypes. This map shows the global distribution and genetic diversity of the major subtypes, known as "clades," of HIV-1.

Occasionally, two viruses of different subtypes can meet in the cell of an infected person and combine their genetic material creating a new hybrid virus (a process similar to sexual reproduction, and sometimes called "viral sex"). Fortunately, many of these new strains do not survive for long, but those that infect more than one person are known as "circulating recombinant forms" or CRFs. For example, the CRF A/B is a mixture of subtypes A and B in the same individual.

SIV and the Origin of AIDS

Scientists today generally agree that simian viruses resembling the human immunodeficiency virus (HIV) originally spread from chimpanzees and sooty mangabeys to people in central Africa. Most AIDS researchers currently believe that HIV is the virus that causes AIDS and that two different strains of a virus affecting monkeys probably combined to create the form of HIV that has spread around the world. A theory put forward by a team from Nottingham University, who performed an extensive genetic analysis of SIV strains, proposes that wild chimps simultaneously infected with two different SIV strains, which had "viral sex" forming a third virus that could be passed on to other chimps and was capable of infecting humans and thereby causing AIDS (*http://news.bbc.co.uk/2/hi/health/2985596.stm*).

In February 1999, a University of Alabama group led by Dr. Beatrice H. Hahn found a type of SIVcpz that was almost identical to HIV-1 from a frozen sample taken from a subgroup of chimpanzees (Pan Troglodytes) once common in west-central Africa. The researchers then collected blood samples from several wild-born baboons and monkeys in Cameroon-17 species altogether. Dr. Hahn and colleagues reported that only chimpanzees in west-central Africa harbor a viral strain that is "truly closely related" to the most lethal AIDS strain. They claimed that this sample proved that chimpanzees were the source of HIV-1 and that the virus at some point crossed species from chimps to humans[11].

Some have theorized that SIVcpz was transferred to humans as a result of administration of

Figure 3-4 HIV is believed to have crossed species from monkeys to humans. (Image © John Austin. Used under license from Shutterstock, Inc.)

the oral polio vaccine (OPV). It has been known for a long time that the attenuated virus used to make the OPV was grown in monkey kidney cells, further fueling this theory. However, in February 2000, the Wistar Institute in Philadelphia, Pennsylvania, who was involved in the distribution of the polio vaccine found some vials that were frozen since the 1950s. Tests performed by three independent laboratories on the 1950s-era polio vaccine samples failed to find any traces of SIV, HIV-1, or DNA indicating that chimpanzee cells were used to prepare the vaccine. It was later confirmed that Asian macaque monkey kidney cells were used for the OPV development and not chimpanzee (*http://www.aidsorigins.com/content/view/121/29/*).

Many now believe that SIVcpz was transferred to humans as a result of chimps being killed and eaten or their blood getting into cuts or wounds of the hunters (**The Hunter Theory**). The earliest known instances of HIV transfer to the human population have also been a point of interest for many. A plasma sample taken in 1959 from an adult male living in what is now the Democratic Republic of Congo tested positive for HIV-1. HIV-1 was also found in tissue samples from an American teenager who died in St. Louis in 1969. HIV was also found in the tissue samples from a Norwegian sailor who died around 1976. Using a new statistical method and one of the world's most powerful computers, scientists at the Los Alamos National Laboratory, led by Dr. Bette Korber, speculate that HIV-1 might have been introduced into humans around the 1930s (*New York Times*, Feb. 2, 2000). HIV-2 is thought to have originated from the SIV in Sooty Mangabeys rather than chimpanzees. In May 2003, a group of Belgian researchers led by Dr. Anne-Mieke Vandamme, concluded that HIV-2 first appeared in humans around 1940 to 1945. All evidence therefore suggests that HIV is a recent disease of humanity.

CHAPTER 4

HIV and the Immune System

Since HIV infection leads to AIDS by destroying the body's defenses, one needs to understand how the immune system functions under normal circumstances in order to appreciate how HIV really works. The immune system is composed of many interdependent cell types that collectively protect the multicellular organism from bacterial, parasitic, fungal, or viral infections and from the growth of tumor cells. Typically the body's immune system vigorously fights infection and disease in a rapid and efficient manner. This is generally accomplished through the numerous cells that make up the major functional components of the system. However, when one part of the system doesn't work correctly, your body becomes vulnerable to many kinds of diseases.

Innate and Adaptive Immune Responses

Humans possess a plethora of immune defensive mechanisms that are employed every day to combat infections. The immune system employs two main types of defense mechanisms when it encounters an **antigen** (any "foreign" substance) that induces the production of immune defense proteins. The first is the **innate** or nonspecific defense, which protects the body from various forms of infectious agents. Innate defenses include physical barriers such as the skin, but also employ the acidity of the gastric acids in our stomach, or lysozyme in our tears as well as phagocytic **macrophages, neutrophils,** and **natural killer (NK) cells** to combat pathogens. These physical barriers are very effective in abrogating the entry of bacterial and viral organisms through destruction of the infectious threat or by physically preventing its attachment to host cells through competition from normal flora. The **adaptive** arm of the immune system is characterized by the specific or directed action of specialized **lymphocytes** and specific antibody production directed against a particular pathogen or pathogen protein (epitope). The basis of the specificity of the adaptive systems lies in the capacity of immune lymphocyte cells to distinguish between proteins produced by the body's own cells ("self" antigens), and proteins produced by invaders or cells under the control of a virus ("non-self" antigen or what is not recognized as the host organism). The cells of the immune system can engulf bacteria, kill parasites or tumor cells, or kill virus-infected cells. All of the cells that function in the immune system are called white blood cells (WBC).

The WBCs of the immune system are produced in the bone marrow and are then carried in the peripheral blood to specialized organs of the immune system, where they are educated to differentiate between self and non-self in preparation to launch immune responses against infections. These cells originate from the lymphatic system organs, which include the thymus, lymph nodes, and bone marrow. As far as scientists have determined, each cell plays a specific role in fighting disease. Every germ that invades the body has unique identification marks on its surface. Some white blood cells, like macrophages, destroy and eat bacteria and damaged cells. B-lymphocytes commonly referred to as B cells mature in the bone marrow and later produce **antibodies** (host defense proteins; see Figure 4-2 on page 26) after differentiation into plasma cells with the help of T-lymphocytes (T cells) called **T-helper cells.** T-lymphocytes mature in a lymphoid organ called the thymus. The antibodies that are produced can neutralize viruses, bacteria, or toxic proteins in the blood and other body fluids, thereby preventing disease.

Central Role of CD4 Cells

Picture the immune system as a group of well-trained Navy Seals in the heat of battle. Because of the terrain in which they find themselves, the members of this outfit cannot directly see or talk to each other, but are organized and directed by a commanding officer who is constantly sending each solider a stream of messages with the GPS coordinates of the enemy as well as his fellow combatants. In the human body, the role of the commanding officer would be filled by the T-helper cell (also known as the "CD4-cell" because of the major receptor on its surface).

The CD4 cell dictates the behavior of the other immune cells almost from the very moment of an infection. The major types of cells controlled by the CD4 cell include the antigen-presenting cells (APCs), which are at the front line of the immune system. These cells are constantly patrolling the tissues and are responsible for identifying foreign entities by recognizing proteins, or antigens, on the surface of an invader, which marks it for death. The APCs then engulf the foreign antigen, break it down, and display portions of it on their surfaces (Figure 4-1). These pieces that are displayed serve as a signal to the CD4 cell that the body is under attack, and it needs to mobilize the other members of the outfit. **B-cells** produce **antibody** molecules (Figure 4-2) that are specifically designed to tackle and "lock" onto antigens on the surface of virus or bacterial particles to prevent them from replicating.

These B cells are activated by chemical signaling proteins called **cytokines,** produced by the CD4 cells, causing the B cells to produce antibodies to the specific threat

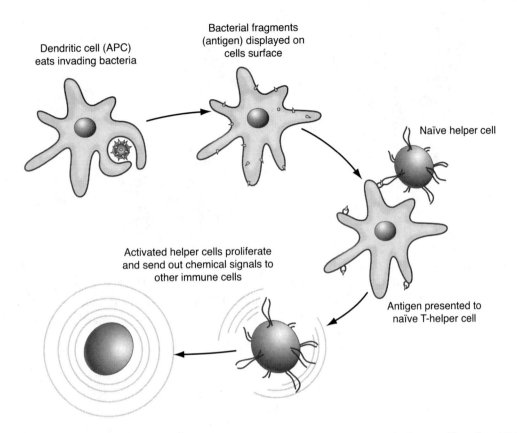

Figure 4-1 Antigen presentation by a dendritic cell (APC) to T-helper cells. The APC engulfs the bacteria and breaks it down into fragments that are presented on the surface of the cell to naïve T-helper cells. These naïve cells become activated, secreting chemical messengers that "call" in other immune cells to the site as well as help to activate B cells and cytotoxic T cells.

(Figure 4-3). The antibody molecules, which are "Y-shaped," attach themselves to the virus or bacteria and prevent them from causing infection.

Antibodies bound to the surface of invaders also serve as a signal for macrophages, which act like scavenger cells that find and engulf any bacteria or virus particles marked by the antibodies. CD4 cells also lead to the activation of **killer T cells** (Figure 4-3) that hunt for viral antigens, which indicate that a body cell has been infected with a virus and turned into a virus-producing machine. The killer T cell, as its name suggests, kills the infected cell to stop it from producing more virus particles. Within a few days, the immune system can eliminate a typical flu or cold virus infection. Once a virus is eliminated, however, the body doesn't forget about it. Some of the B cells and

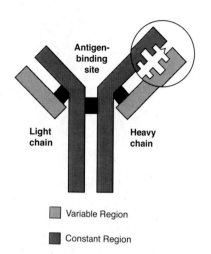

Variable Region

Constant Region

Figure 4-2 Antibody molecule. An antibody is made up of two heavy chains and two light chains. The variable region, which differs from one antibody to the next, allows an antibody to recognize its matching antigen. (Source: NIH)

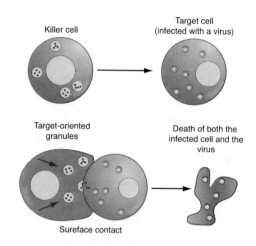

Figure 4-3 Killer T cells directly attack other cells carrying certain foreign or abnormal molecules on their surfaces. (Source: *http://en.wikipedia. org/wiki/ Immune_response*)

T cells that were mobilized for the fight become memory cells that stay in the body for many years. If the same virus comes back, those cells will quickly find it, clonally expand to increase the number of responders and eliminate it, before you ever become sick. That is why some diseases, like the measles, only strike once in a lifetime.

How Does HIV Attack the Immune System?

HIV, like the other members of the lentivirus family of retroviruses, is unique among viruses because it directly attacks the CD4 helper T cell, the very heart of the immune system. By targeting this irreplaceable cell, the virus interrupts the entire immune response, leaving the host vulnerable to any infectious agent. Once inside the body, HIV hunts down, attaches to, and penetrates the CD4 cell. Remember that the other cells need the CD4 cell as their commander to tell them what they need to do to rid the body of the invader (See Figure 4-4). HIV, however, hijacks the CD4 cell and uses its machinery to make virus particles. Although some immune cells, including NK cells and cytotoxic T-lymphocytes, recognize and destroy HIV particles, the virus readily attaches to the CD4 receptor of the helper T cells and enter the cell where it starts a productive infection. Since cells of the immune system are constantly circulating in and out of the lymphoid organs, the HIV-infected cell will soon carry the virus into

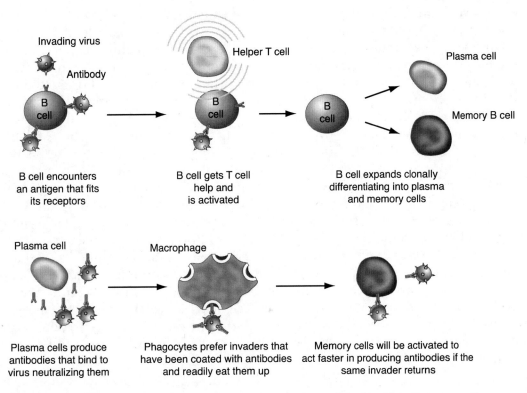

Invading virus

Antibody

B cell

B cell encounters
an antigen that fits
its receptors

Helper T cell

B cell

B cell gets T cell
help and
is activated

B cell

Plasma cell

Memory B cell

B cell expands clonally
differentiating into plasma
and memory cells

Plasma cell

Plasma cells produce
antibodies that bind to
virus neutralizing them

Macrophage

Phagocytes prefer invaders that
have been coated with antibodies
and readily eat them up

Memory cells will be activated to
act faster in producing antibodies if the
same invader returns

Figure 4-4 B cell activation by specific antigen and T cell help. This figure outlines the very important role played by the T-helper cells in the immune response. Without T cell help, the B cell does not expand and differentiate.

areas of the lymphatic system, where CD4 cells are most abundant. Once there, HIV begins replicating and infecting CD4 cells more rapidly, replicating itself billions of times each day.

Cellular Receptors and Entry of HIV

As illustrated in Figure 2-2, HIV resembles a sphere with spiky proteins all over its surface. In order to infect a CD4 cell, one of those proteins, known as gp120, binds to a receptor protein, CD4, on the T cell's surface. A second protein, called gp41, then locks onto a second receptor protein, CCR5 or CXCR4 as previously discussed. Once both connections are made, the virus can penetrate the outer wall of the cell and insert its viral proteins and RNA. Once inside the cell, the virus hijacks the cellular machinery, transforming it into an HIV factory. As previously described under HIV replication, several enzymes are employed by the virus to transcribe viral RNA into DNA, to integrate the viral DNA into the cellular DNA, and to activate the provirus. After the copies are made, they gather at the cell's surface and form buds before finally disconnecting from the host cell. Once they're free, an enzyme called protease cuts the new proteins and RNA into small, usable pieces. The new virions are now mature, and ready to do battle with the body. This process of infection and viral budding destroys the CD4 cells, slowly depleting their numbers (Figure 4-5). Many people experience flu-like symptoms, sometimes with a rash, two to three weeks after being infected. This is the body reacting to the invading HIV and mounting a strong immune response. At this point the body is producing up to a billion helper and killer T cells a day. The constant stimulation of the immune system leads to activation of CD4 cells that are already infected, leading to an increase in the production of new viruses. Those viruses in turn go out and infect more of the newly created CD4 cells. At this point, there is a large enough viral load in the blood that HIV can be diagnosed through a blood test.

Soon, the extensive damage to these T-helper cells takes its toll and the total number of these cells begins to fall steadily. As destruction continues, viral particles are released into the bloodstream and the viral load increases steadily. Over time, the antibody levels will fall since there is no more T cell help for the HIV-specific B cells to continue producing these defense proteins. The reduction in CD4 cell count with a concomitant reduction in protective antibodies signals the progression of HIV infection and signs of immunodeficiency soon become visible.

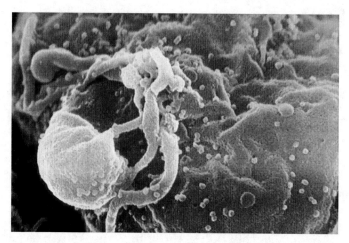

Figure 4-5 Scanning electron micrograph of HIV-1 budding from cultured lymphocyte. Multiple round bumps on the cell surface represent sites of assembly and budding of virus particles. (Source: Content providers: CDC/ C. Goldsmith, P. Feorino, E. L. Palmer, W. R. McManus. *http://phil.cdc.gov/PHIL_Images/10000/10000.tif*)

CHAPTER 5

Progression of HIV Infection to AIDS

The battle between this resilient retrovirus and the immune system cells can continue to rage in the body for up to 10 years, since the two entities have such a large supply of ammunition. During this time, the HIV-infected person might feel completely healthy. The infected individual can, however, spread HIV to anyone with whom he or she exchanges bodily fluids, such as blood, vaginal fluid, semen, or breast milk, because the virus is active. When an infected cell starts producing HIV proteins, the HIV envelope proteins migrate to the cell membrane and are displayed on the cell surface, just as if they were poking out of an HIV particle. The infected cell now looks like an HIV particle and is able to bind to other CD4 positive cells. Therefore, an infected T-helper cell can join with a healthy, uninfected T-helper cell in a similar manner as the HIV particle originally attached to and entered into the host cell. This is repeated until eventually you have one large HIV-infected CD4+ cell with as many as 50 nuclei called a Syncytium (Figure 5-1). Syncytia do not always form in HIV-infected patients (about 50% of people after five years), however, when it does, it serves as a route for rapid CD4 cell depletion since these bound cells cannot function in immune protection. Individuals with syncytia progress to AIDS more rapidly and are more likely to have neurological involvement.

A major reason that HIV is unique is the fact that despite the body's aggressive immune responses, which are sufficient to clear most viral infections, some HIV invariably escapes. This is due in large part to the high rate of mutations that occur during the process of HIV replication. Even when the virus does not escape the immune system by mutating, the body's top soldiers in the fight against HIV—subsets of killer T cells that recognize HIV—may be depleted or otherwise become dysfunctional. Eventually, HIV gains the upper hand and begins to kill off more CD4 cells than the body can produce. Without the CD4 cells, the B cells and killer T cells lack instruction about where to go and what to look for, and the body's immune response becomes less and less effective. The viral load in the plasma quickly rises and the amount of

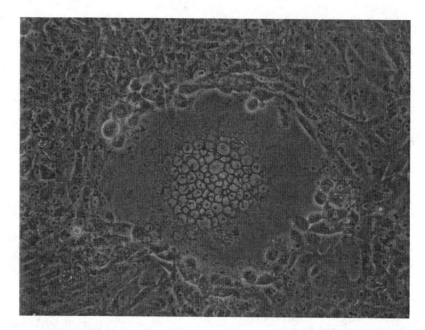

Figure 5-1 Syncytium "forming"-type cytopathic effect induced by the virus infection. Phase-contrast microscopic image of 100-fold magnification. (Source: Ytombe's file. Permission: GDFL)

antibodies produced by the B cells fall. This leaves the patient with no real protection against the weakest of pathogens.

Stages of HIV Infection

In addition to the serious opportunistic infections that characterize the AIDS stage of HIV infection, there may be several other signs and symptoms of the disease at various stages of the infection process. It is not unusual for people with AIDS to have systemic symptoms of infection, including fevers, chills, night sweats, swollen lymph glands, generalized weakness, and severe weight loss. Using the CD4 cell counts, HIV antibody levels, and the opportunistic infections associated with HIV infection, the entire infection process can generally be broken down into four distinct stages: **primary infection, clinically asymptomatic stage, symptomatic HIV infection,** and **progression from HIV to AIDS.**

The **primary stage** of infection lasts for only a few weeks and is often accompanied by a short flu-like illness. Almost 20% of those infected feel sick enough to consult a doctor at this stage of the illness. At the beginning of the primary stage the viral load increases rapidly, inducing an immune response. At this point, an antibody test would be negative because the immune system needs time to respond and make HIV-specific antibodies. As antibody production increases (seroconversion) and killer T cells start responding to the virus, the viral load begins to fall because the antibodies and T cells keep them in check. The period between infection and seroconversion is referred to as the **window period** because all antibody tests will be negative at this stage. However, if nucleic acid tests are done at this time they will come up positive for the presence of viral RNA.

The **clinically asymptomatic stage** of HIV infection is very characteristic of lentiviruses of which HIV is a family member. This period can last up to 10 years and may be quite uneventful, being free from major symptoms, although there may be swollen glands.

The level of HIV particles in the peripheral blood falls to very low levels, but people remain infectious throughout this stage and HIV antibodies are detectable in the blood. It should be noted that HIV is by no means dormant during this stage, but is very active in the lymph nodes and other areas where CD4 cells are plentiful. Antibody tests will be positive at this stage and CD4 cell counts will start to fall.

As time progresses the lymph nodes and other tissues become damaged because of the years of active HIV infection. The virus mutates and becomes more virulent, leading to increased T helper cell destruction. Despite its best efforts, the body fails to keep up with replacing the T helper cells that are lost, and the CD4 levels fall precipitously as reflected in peripheral blood counts. As the immune system continues to fail, symptoms begin to manifest. Initially many of the symptoms are mild, but as the immune system deteriorates the symptoms worsen. This is the **symptomatic stage** of HIV infection, which is mainly caused by the emergence of opportunistic infections and cancers that the immune system would normally prevent. These can occur in almost all the body systems. The most common opportunistic infections are discussed below.

Progression of HIV to AIDS is the last stage of the disease. At this stage the body becomes overwhelmed with opportunistic infections as more CD4 cells are lost. The CD4 cell count drops below 200 and the patient feels very ill. Most of the day is spent sleeping or resting and there might be profound weight loss, leaving the patient looking emaciated. Studies suggest that HIV also destroys precursor cells that mature to have special immune functions, as well as the microenvironment of the bone marrow and the thymus that is needed for developing such cells. These organs probably lose the ability to regenerate, further compounding the suppression of the immune system.

When Is a Person Diagnosed with AIDS?

Untreated HIV disease is characterized by a gradual deterioration of immune function. Two to four weeks after exposure to the virus, up to 70% of HIV-infected people suffer flu-like symptoms related to the acute infection. With the persistent depletion of CD4 cells, the immune system loses its ability to protect the infected patient from infections that would ordinarily be easily cleared. A healthy, uninfected person usually has 1,000 to 1,500 CD4+ T cells per cubic millimeter (mm^3) of blood. If an individual is infected with HIV and is not taking antiretroviral medication, the number of these cells in a person's blood progressively declines (Figure 5-2).

When the CD4+ T cell count falls below 200/mm^3, the infected person will become very susceptible to several opportunistic infections and cancers that are characteristic of immunodeficiency or AIDS, the end stage of HIV infection.

AIDS patients often suffer from and later succumb to infections of the lungs, intestinal tract, brain, eyes, and other organs. This is typically accompanied by debilitating weight loss, diarrhea, neurological disorders, and certain rare cancers such as Kaposi's sarcoma and certain types of lymphomas. There is widespread HIV-mediated destruction of the lymph nodes as well as the spleen and other organs of the immune system. This leads to further immunosuppression seen in people with AIDS.

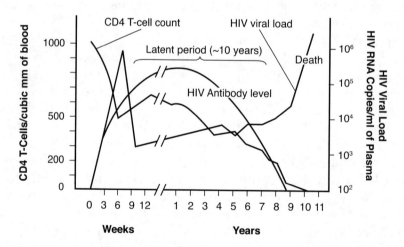

Figure 5-2 Progression of HIV infection to AIDS. Note that as the antibody titer increases, the viral load is decreased. As the CD4 count falls, so does the antibody level and the viral load again soars.

Immunosuppression by HIV is confirmed by the fact that medicines, which interfere with the HIV life cycle, result in the preservation of CD4+ T cells and immune function as well as delay clinical illness. HIV infection leading to AIDS, however, is not uniformly manifested in all individuals. A relatively small proportion of individuals infected with HIV rapidly develop AIDS and die within months after primary infection. At the same time, approximately 5% of HIV-infected individuals exhibit no signs of disease progression even after 15 or more years. Antiretroviral drugs can significantly prolong the time between initial HIV infection and the onset of AIDS. Modern combination therapy is very effective and, theoretically, someone with HIV can live for a very long time before progressing to AIDS. Unfortunately, these potentially lifesaving medicines are not widely available and are quite expensive. Therefore, millions of AIDS patients in many poor countries around the world, who cannot access or afford the medication, continue to die. It is believed that various host factors such as the virulence level of the individual HIV strain, age of the person at the time of infection, genetic differences among individuals, immune status, as well as co-infection with other microbes may determine the rate and severity of HIV disease expression from person to person.

AIDS Opportunistic Infections

Each of us carries in our bodies an array of germs—viruses, bacteria, fungi, and protozoa—that exist in a balanced state with our immune system. When our immune system is working well, it controls these germs. However, when the immune system becomes compromised by HIV, other bacterial and viral diseases, or even by some medications, these germs can grow out of control and cause health problems. HIV does not kill its victims directly; instead, it weakens the body's ability to fight diseases by destroying our T helper cells as explained earlier. Consequently, infections that rarely occur in individuals with a properly functioning immune system can prove deadly to HIV-infected individuals. These infections are collectively termed **opportunistic infections** because they take advantage of the opportunity offered by a weakened immune system to grow in the host at a higher level than they are accustomed to being present, and must be treated if the patient hopes to survive. The CDC has developed a list of opportunistic infections signifying that an individual who is HIV positive has progressed to AIDS (*http://www.hivandhepatitis.com/hiv_ois_list.html*). When an HIV-positive person has both a low CD4 count and what doctors call an "AIDS-defining illness" or "opportunistic infection," that person has AIDS. Early symptoms of worsening HIV infection include night sweats, weight loss, fatigue, and swollen lymph nodes.

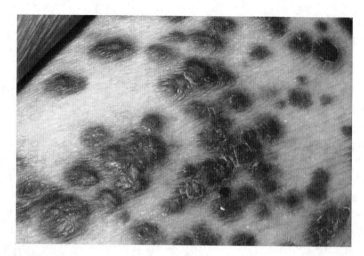

Figure 5-3 The purplish lesions of Kaposi's sarcoma in an AIDS patient. (Source: National Cancer Institute)

Common Opportunistic AIDS Infections

Some of the most common opportunistic infections affecting HIV/AIDS patients are:

1. **Kaposi's sarcoma (KS)**—A type of skin cancer that presents with purple lesions all over the body (Figure 5-3). Before the advent of AIDS, KS was seen only rarely in older men, mostly of eastern European descent. Although KS was one of the defining signs of the disease in the early years of AIDS, with current treatment regimens, KS is less commonly seen than it was during the 1980s and early 1990s.

2. **Pneumocystis carinii pneumonia (PCP)**—A rare form of pneumonia typically seen in people who are severely immunosuppressed. The main symptoms are shortness of breath and difficulty breathing, as the lungs fill up with fluid (Figure 5-4). Treatment requires heavy doses of antibiotics combined with oxygen therapy. PCP continues to be a common infection in people with AIDS, and many suffer several bouts of this serious infection before dying.

3. **Toxoplasmosis**—An infection caused by a parasite, *Toxoplasma gondii*, that infects the brain and can cause brain damage and blindness. The infection affects 10% to 20% of people in North America by the time they are adults, but the immune system normally keeps it under control. Before AIDS, toxoplasmosis was consid-

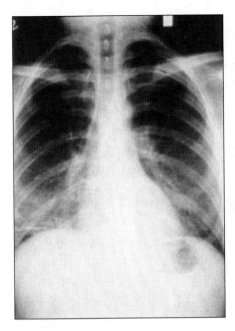

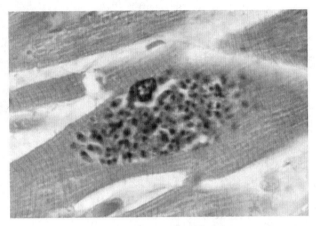

Figure 5-4 Lung X-ray of a patient diagnosed with AIDS shows PCP infection. There is increased white (opacity) in the lower lungs on both sides, characteristic of *Pneumocystis* pneumonia. (Source: NIH, CDC Website)

Figure 5-5 Toxoplasmosis of the heart in AIDS. Histopathology of active toxoplasmosis of myocardium (heart muscle). Numerous tachyzoites (asexual stage of rapid growth) of *Toxoplasma gondii* are visible within a pseudocyst in a myocyte (Muscle cell). Parasite. CDC/ Dr. Edwin P. Ewing, Jr. (Source: *http://phil.cdc.gov/ PHIL_Images/ 966/966.tif*)

ered a minor danger only for pregnant women when changing a cat litter box or working in an area contaminated with cat feces. Infection in the unborn child early in pregnancy can result in miscarriage, poor growth, early delivery or stillbirth. If a child is born with toxoplasmosis, he or she can experience eye problems, hydrocephalus (water on the brain), convulsions, or mental disabilities. Toxoplasmosis has become a serious infection in AIDS patients and may infect the eyes, brain, and heart muscle cells (Figure 5-5).

4. **Candidiasis**—A common opportunistic infection in people with HIV. It is an infection caused by a common type of yeast (fungus) found in most people's bodies. A healthy immune system keeps it under control. However, at the point

of immunosuppression, candidiasis runs rampant, infecting the mouth, throat, windpipe, skin, or vagina. Candidiasis can occur months or years before other, more serious opportunistic infections. It appears most frequently in men as a white coating of the mouth (Figure 5-6) and causes burning, a bad taste, and lack of appetite. Women often experience this infection as chronic vaginal candidiasis that does not respond to treatment. This infection can infect organs and tissues throughout the body becoming life threatening.

5. **Cytomegalovirus (CMV)**—Another virus that is common in adults, but is normally held in check by a healthy immune system. Between 50% and 85% of the U.S. population tests positive for CMV by the time they are 40 years old. CMV is a member of the herpes family of viruses and currently one of the leading causes of blindness and death in people with AIDS. CMV can infect any part of the body, but in people with AIDS, it generally attacks the eyes and lungs (Figure 5-7).

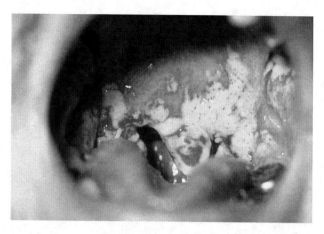

Figure 5-6 Oral thrush or thrush. Note the opaque areas on the gum lining outlined by the circle. (Source: CDC *http://phil.cdc.gov/ PHIL_Images/02112002/00018/PHIL_1217.tif*)

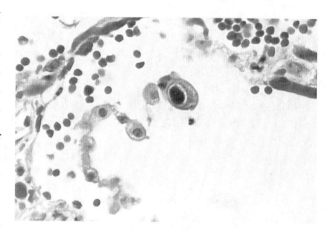

Figure 5-7 Lung histopathology of CMV. The central cell displays the dramatically enlarged intra-nuclear inclusion, characteristic of CMV. (Source: CDC *http://phil.cdc.gov/PHIL_Images/ 958/958.tif*)

6. **Tuberculosis (TB)**—TB is unique among HIV-related infections because it can infect people who are not infected with HIV or are immunocompetent. It is a respiratory pathogen that was on the decrease in most parts of the world until the 1980s when the AIDS epidemic started. TB is transmitted through a respiratory route but is treatable once identified. This is one of few infections that may occur in early-stage HIV disease. However, multi-drug resistance is a potentially serious problem. Although the incidence of HIV-associated TB has declined over the past ten years because of drug therapy and other improved practices in Western countries, it

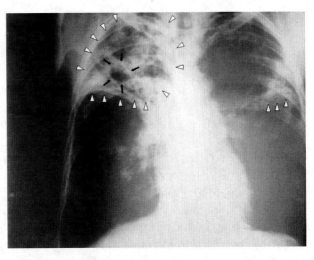

Figure 5-8 An anteroposterior X-ray of a patient diagnosed with advanced bilateral pulmonary tuberculosis. This AP X-ray of the chest reveals the presence of bilateral pulmonary infiltrate (white triangles), and "caving formation" (black arrows) present in the right apical region. The diagnosis is advanced tuberculosis. (Source: CDC)

remains a serious problem in developing countries where HIV is most prevalent. In early-stage HIV infection (CD4 count >300 cells per μL), TB typically presents as a pulmonary disease (Figure 5-8). However, in advanced HIV infection, TB often presents with systemic disease affecting other extra pulmonary organs. Symptoms are usually not localized to one particular site and may involve infection of the bone marrow, urinary and gastrointestinal tracts, liver, regional lymph nodes, and the central nervous system.

AIDS Dementia Complex

The AIDS dementia complex (ADC), also known as HIV-1-associated dementia or HIV-associated cognitive/motor complex, has been estimated to affect up to one-third of adults and one-half of all children with AIDS. ADC is one of the most common and clinically important central nervous system (CNS) complications of late HIV-1 infection. It develops principally in the late stage of HIV-1 infection and associated severe immunosuppression. ADC might be one of the only AIDS-related complexes directly caused by the HIV virus. Those with ADC have HIV-infected macrophages in the brain. That means HIV is actively infecting brain cells. There are several theories about how this might come about. The most common explanation currently is that HIV enters the brain via HIV-infected monocytes, which then differentiate into macrophages. The virus replicates in these cells and can then, in theory, infect other cells in the immediate surroundings; macrophages and microglial cells are the most common. These infected cells are thought to secrete chemical messengers and inflammatory proteins that lead to neurotoxicity. The term *dementia* is used by neurologists to describe a clinical syndrome composed of memory loss, decreased mental concentration, and the loss of other intellectual functions because of progressive disease in the brain. There may be motor signs, such as weakness, lack of coordination, and unsteady gait. Other symptoms may include loss of interest in one's surroundings, as well as severe mobility problems. Prior to effective antiretroviral therapy, ADC occurred in more than 60% of patients who developed AIDS. With the use of highly active antiretroviral therapy (HAART), the incidence has declined to about 15% to 25%. ADC is often treated with antipsychotics, antidepressants, psychostimulants, antimanics, and anticonvulsants. These drugs do not treat the underlying cause of ADC, or even stop its progression. However, they may ease some of its symptoms.

CHAPTER 6

Transmission of HIV

With the absolute devastation that AIDS has brought upon the world's population, you might be surprised to learn that compared to other viruses such as hepatitis A, which can remain active for an extended period of time outside the body, HIV is a very fragile virus. The fact is, outside the body, HIV cannot withstand the many environmental pressures that most organisms have to contend with in order to survive. Factors such as heat, desiccation or drying, and various chemicals prove detrimental to the virus. Therefore, HIV is spread through direct contact with the body fluids of an infected person and cannot be contracted by casual contact. Although HIV has been found in tears, sweat, and saliva, the number of viral particles in these fluids is very low. Health officials maintain that the numbers in these fluids are too low to result in transmission of a productive infection.

Major Routes of HIV Transmission

The main routes of HIV infection are: (1) penetrative unprotected sexual intercourse with someone who is infected with the virus, (2) transfusion of contaminated blood or injection of contaminated blood products, (3) organ transplants or skin grafts taken from someone who is infected, (4) artificial insemination with infected semen, (5) transmission from an infected mother to her baby during pregnancy, at the time of birth or through breast-feeding, and (6) sharing of an unsterilized needle that was previously used by someone who is infected. Infection through contaminated blood and blood products is currently very rare since blood and blood products are routinely screened for HIV antibodies or nucleic acid material. There has been only one documented instance of a health care worker transmitting HIV to patients in the United States. This was an unfortunate case where one infected dentist transmitted HIV to six of his patients during surgery[8].

Worldwide, the virus is acquired most commonly through infected semen and blood. Sexually transmitted HIV worldwide is more prevalent in heterosexual relationships. In the United States, however, homosexuals and injection-drug users

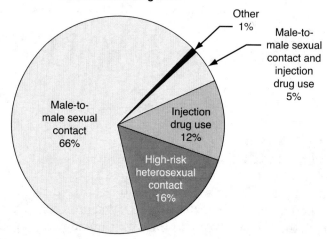

Figure 6-1 HIV transmission by categories for male adults and adolescents in the United States during 2006. MSM accounted for over two-thirds of all male transmissions. (Source: This Graph was created based on HIV/AIDS statistics from the CDC. *http://www.cdc.gov/hiv/resources/factsheets/us.htm*)

account for the highest proportion of HIV-infected persons. HIV and AIDS have had a tremendous impact on men who have sex with men (MSM) in the United States. Although only about 5% to 7% of men in the United States identify themselves as MSM, this group accounted for approximately two-thirds of all HIV infections among men in 2006 (See Figure 6-1)[12].

Having another sexually transmitted infection (STI) may also impact one's susceptibility to HIV infection. If an HIV-infected person has another STI, the presence of this STI can increase the risk of HIV transmission. If an HIV-negative individual has an STI, it can also increase their risk of being infected with HIV. This may be because most STIs (e.g., syphilis or herpes) cause the formation of genital warts or ulcers, which could bleed or which can allow the transfer of infected body fluids directly into the bloodstream. Intact, healthy skin is an excellent barrier against HIV, other viruses, and bacteria. Infection with other STIs such as *Chlamydia* or gonorrhea may cause localized inflammation leading to an influx of immune cells in the genital area, thus making HIV transmission much more likely. HIV transmission, however, is more likely in individuals with genital ulcers than in those with an STI that does not cause ulcers. Using condoms during sex is the best way to prevent the sexual transmission of diseases, including HIV.

High-Risk Behaviors

The CDC estimates that approximately 4 to 5 million Americans participate in activities that put them in danger of coming in contact with infected blood or other body fluids. Anyone who participates in these high-risk behaviors is at increased risk of contracting HIV. These activities include sharing needles or syringes; having sexual contact, including oral, with an infected person without using a condom; or having sexual contact with someone whose HIV status is unknown. HIV is frequently spread among injection-drug users by the sharing of needles or syringes contaminated with very small quantities of blood from someone infected with the virus (Figure 6-2). In many parts of the world, often because it is illegal to possess them, injecting equipment or works are shared. A very small amount of blood can transmit HIV, and can be injected directly into the bloodstream along with the drugs.

Female prostitutes often have 200–300 sexual partners per year and are therefore at a much higher risk for exposure to HIV and AIDS than the vast majority of heterosexuals. While many HIV/AIDS dissidents argue that prostitutes should have a much higher overall prevalence of the infection, the fact is that most of these professional sex workers mandate that their customers use condoms or have other means of barrier protection. The nation of Senegal has learned a hard lesson about the folly of the reasoning that legalized/regulated prostitution will lower the incidence of sexually transmitted infections. Senegal has tolerated prostitution among woman over 21 years of age since 1969. While Senegal has one of Africa's lowest overall infection rates at less than 1%, this disguises a dangerous rise among vulnerable groups like sex workers. HIV prevalence was not only much higher, but growing among this group. According to **Enda Third World,** an international NGO based in Dakar, today, HIV infection among legal sex workers in the capital, Dakar, has risen sharply to 21% compared with 1% two decades ago.

Male prostitution has been shown to pose just as high a risk of transmission. The fact is that high rates of HIV have been consistently found among individuals who sell sex in many different countries and cultures. Even in areas where HIV prevalence is generally low, the rate among sex workers is usually higher than the rate found among the general

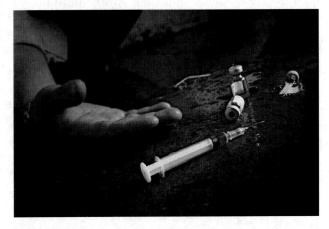

Figure 6-2 Injection drug use is a major route of HIV transmission. (Image © Wessel du Plooy. Used under license from Shutterstock, Inc.)

adult population. In some places, sex workers commonly use drugs and share needles and this overlap between sex work and *injecting drug use* is linked to growing HIV epidemics in a number of countries, such as China, Indonesia, Kazakhstan, Ukraine, Uzbekistan, and Vietnam.

HIV Transmission in MSM

Figure 6-3 Preventing HIV infections among sex trade workers has been proven to be an instrumental part of the fight against AIDS in many countries. (Image © Igor Balasanov. Used under license from Shutterstock, Inc.)

As mentioned briefly earlier, MSM are at increased risk for infection with HIV (See Figure 6-1 and Figure 6-4). This is a direct result of the ease with which the virus gets into the bloodstream through infected semen during anal sex. It turns out that rectal tissue is particularly easily disrupted or torn and the trauma of the sexual act leads to this type of tearing. While it is not entirely apparent why there is an increase in unprotected anal intercourse, research evidence suggests several factors including improvements in HIV treatment leading to more healthy-looking men who are able to have sex without their partners knowing that they are indeed HIV positive.

Seeking sex partners on the Internet is another factor that increases one's chances of ending up with an infected partner. The Internet has the potential to normalize certain high-risk behaviors by making others aware of these behaviors and creating new connections between the men who engage in them.

The use of alcohol and illegal drugs continues to be prevalent among some MSM and is linked to HIV and STI risk. Data show that there is a tendency toward risky sexual behaviors while under the influence of various drugs. These behaviors include sharing needles or other injection equipment. Maybe most important among these risk factors is a failure to maintain prevention practices such as condom use[13]. The proper and consistent use of latex or polyurethane (a type of plastic) condoms when engaging in sexual intercourse, whether vaginal, anal, or oral, can significantly reduce a person's risk of acquiring or transmitting sexually transmitted diseases, including HIV infection.

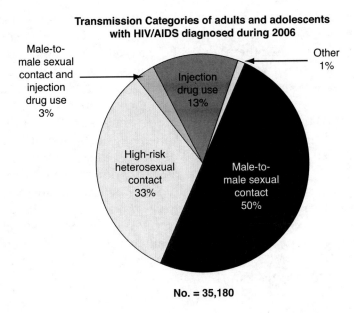

Figure 6-4 Diagram showing HIV transmission categories of all adolescents and adults in the United States *in 2006*. The largest estimated proportion of HIV/AIDS diagnoses among adults and adolescents were men who have sex with men (MSM), followed by persons infected through high-risk heterosexual contact. (Source: The diagram was created from CDC data. *http://www.cdc.gov/hiv/resources/factsheets/us.htm*)

Bug Chasers and HIV Infection

With the devastation that HIV has caused over the past 26 years, it would be logical to conclude that this is the last disease that anyone would want to end up with. It might therefore surprise the average citizen that there are a number of men who are purposely trying to become HIV positive! A **bug chaser** is a gay man who deliberately attempts to contract HIV by having unprotected sex with a man or a group of men who are known to have the virus.

"In private sex clubs across the U.S. men gather for a chance to participate in what is called Russian Roulette. Ten men are invited, nine are HIV-negative, and one is HIV positive. The men have agreed to not speak about AIDS or HIV. They participate in as many unsafe sexual encounters with each other as possible, thus increasing their chances to receive 'the bug.' These are the men known as 'Bug Chasers'," writes author Daniel Hill (*http://www.alternativesmagazine.com/15/hill.html*). The HIV-positive gay man who deliberately transmits the virus to bug chasers is called the "gift giver." Some psychologists theorize that participation in these crazy bug parties is actually an anxiety disorder where the uninfected individuals fear getting HIV so greatly that they would rather contract it and free themselves of the anxiety of living in fear! Bug chasing has been viewed largely with disdain from the gay community and is seen as a self-destructive activity. Leaders of the gay community at large are concerned that the behaviors of bug chasers may contribute to a public perception that the practice is common or encouraged by all gay people, and would thus cause further ill-will toward them (*http://en.wikipedia.org/wiki/Bugchaser*). Daniel Hill further reports that there are men who, once infected, feel like they finally "belong"; they are now part of the gay community. Some find relief in knowing that now they don't have to worry about getting infected any more, since the deed is done. Some believe the myth that HIV is a chronic manageable disease and that the new drugs promise them a long and healthy life. Some couples see infection as the deepest level of intimacy. Whatever the reason for this unusual longing to be infected with a deadly virus, I think it is safe to say that most gay men with HIV do not want to pass HIV on, and most gay men who do not have HIV do not want to get infected.

Women and HIV Transmission Risk

Based on data collected to date, gender inequalities as well as biological factors make women and girls especially vulnerable to HIV and to the overall impact of AIDS. Biologically, women are more susceptible to HIV infection than men and are more likely to become infected in any given heterosexual encounter. This is due primarily to the following factors: (1) women have a greater surface area of mucous membrane exposed during sex compared to their male companions; (2) men transfer a greater quantity of fluids to women during sexual intercourse; (3) male ejaculate fluids contains a higher viral content; and lastly (4) the frequency of micro-tears that can occur in vaginal (or rectal) tissue from sexual penetration leave women vulnerable to direct introduction of HIV into the blood circulation.

Gender biases and norms often increase the HIV transmission risk in women and girls. In many areas of the word, cultural norms allow men to have more sexual partners than women, and even encourage older men to have sexual relations with much younger women. In certain countries this practice has contributed to higher infection rates among young women (15–24 years) compared to young men. Often, women may want to ask their partners to use condoms or to abstain from sex altogether, but are too afraid or lack the power and support to do so. This often leads to men passing HIV on to many partners, often in the same community.

Many women also experience sexual, emotional, and physical violence at some point in their lives. In many of these situations women are forced to have sex which greatly contributes to HIV transmission due to the increase of vaginal tears and lacerations resulting. The threat of violence also prevents women from negotiating safer sexual practices, sharing their HIV status if the results are positive, or accessing the necessary treatment. These major challenges are further compounded by the fact that millions of women worldwide do not have access to educational opportunities, resulting in a lack of economic security, which further contributes to their vulnerability to HIV.

Vertical Transmission

There is a 26% chance of HIV transmission from an infected, untreated pregnant mother to her unborn child. While the percentage of HIV-positive women who give birth to HIV-positive infants has declined "dramatically" in developed countries, due largely to the use of antiretroviral drugs such as **AZT** (Azidothymidine), a lack of prenatal care increases the likelihood that a woman will transmit the virus to her infant. In countries like Uganda, however, mother-to-child transmission is the second most common method of HIV transmission after sexual intercourse. Those women who live in areas where antiretroviral drugs are not readily available transmit the virus to their unborn child or at the time of birth. In sub-Saharan Africa, 40% of all HIV/AIDS cases result from mother-to-child transmission. Recent evidence confirms that antiretroviral drugs are effective in the prevention of vertical transmission of HIV. The treatment regimen, however, is expensive and many people in developing countries cannot afford the approximately $53 U.S. for both the mother and child to be treated with an AZT regimen.

Other Possible Modes of HIV Transmission

There is a low risk of HIV transmission when having a tattoo or body piercing if the instruments contaminated with blood are not sterilized between clients. It is therefore important to determine if people who carry out body piercing or tattoos are following the proper procedures at all times. There are established protocols termed "universal precautions," which govern the handling of body fluids or instruments that might be contaminated with these fluids, designed to prevent the transmission of blood-borne infections such as HIV and hepatitis B. Under the "universal precaution" principle, blood and body fluids from all persons should be considered as being infected with HIV, regardless of the known or supposed status of the person. These guidelines may be found on the following World Health Organization (WHO) site: *http://www.who.int/hiv/topics/precautions/universal/en/*. While there have been unconfirmed stories of HIV transmission at barber shops, this constitutes a very low-risk scenario. Your risk of getting infected with HIV at the barber shop is very low and would only happen if the skin was cut and infected blood got into that wound.

HIV Infection in Health Care Workers

Health care workers may become infected through accidental injuries from needles and other sharp objects that may be contaminated with HIV. The risk is, however, extremely low. It has been estimated that the risk of infection from a needle-stick injury is less than 1%. In the United States, there have only been 56 documented cases of occupational HIV transmissions. All health care workers, including dentists, are required to follow the universal precaution guidelines, making the risk of getting HIV from health care professionals very low.

While it is true that in the past many people became infected with HIV through blood transfusions and blood products which were contaminated with the virus, these products no longer pose a threat, as whole blood donations and blood products are routinely tested for the presence of HIV-1 and 2. In the early 1980s, many hemophiliacs were accidentally infected with HIV. Hemophiliacs typically lack a critical clotting factor (Factor VIII) needed to prevent severe and possibly fatal bleeding with even a small cut. This factor VIII was made from the batch plasma samples of up to 20,000 pooled human donations. Since no one knew about HIV, it was not assayed for and ended up in many batches of the product that these patients inject directly into their veins daily to survive. There are reports that even after some manufactures realized that the products might be HIV infected, they did not immediately pull them from the shelves. Many people got infected this way and also infected their spouses (*http://www.thebody.com/whatis/hemophilia.html*). Today these products are tightly regulated and are routinely tested for HIV.

You Cannot Get HIV From That!

There are probably more myths about ways in which HIV can be transmitted than those about "Big Foot." While this section will in no way serve as the HIV myth buster narrative, I believe it is important to address a few of the more common ones. While HIV has been found in saliva and tears in very small quantities, contact with saliva, tears, or sweat has never been shown to result in transmission of HIV. The CDC, however, advises against deep French kissing with someone who is HIV positive, since ulcers or abrasions from brushing in the mouth could allow enough viral particles to be transferred. Stories about some bitter HIV-positive person planting syringes of their blood and HIV or HIV-positive people getting stuck by needles in phone booths, coin returns, movie theater seats, gas pump handles, and other places are highly unlikely. This is because, as stated earlier, the virus is very fragile and unless these needles are attached to syringes with fresh infected blood they would not be effective in transmitting the live virus.

It is also not possible to get HIV from mosquito bites. When obtaining a blood meal from someone, mosquitoes do not inject blood from any previous individuals that they might have bitten. The mosquito only injects saliva, which acts as a lubricant to enable it to better obtain its meal. Moreover, if mosquitoes could transmit HIV infection, many more young children and preadolescents would have been diagnosed with AIDS, since they are more likely to be exposed to mosquito bites. After much research it appears that HIV is not transmitted by insects. The bottom line is that the virus is not transmitted through normal everyday activities such as shaking hands, toilet seats, swimming pools, sharing cutlery, having contact with animals, regular kissing, sneezes, or coughs. Transmission requires exchange of body fluids from an infected person.

CHAPTER 7

Prevalence of HIV/AIDS

"In June of 1981 we saw a young gay man with the most devastating immune deficiency we had ever seen. We said, 'We don't know what this is, but we hope we don't ever see another case like it again'" remarked Dr. Samuel Broder, then of the National Cancer Institute (WHO, 1994). Unfortunately, Dr. Broder's wish was not granted. Every day, over 6,800 persons become infected with HIV and over 5,700 persons die from AIDS, mostly because of inadequate access to HIV prevention and treatment services. Every 13 seconds someone contracts HIV and every 15 seconds another person dies from AIDS! An estimated 33.2 million [30.6-36.1 million] people worldwide were living with HIV at the end of 2007 (Table 7-1). An estimated 2.5 million [1.4–3.6 million] became newly infected with HIV and an estimated 2.1 million [1.9–2.4 million] lost their lives to AIDS, despite recent improvements in access to antiretroviral treatment.

Overall, the HIV incidence rate (the proportion of people who have become infected with HIV) is believed to have peaked in the late 1990s and to have stabilized subsequently, notwithstanding increasing incidence in several countries. Ninety-six percent of people with HIV live in the developing world, most in sub-Saharan Africa.

Table 7-1 Official global estimates of the HIV/AIDS epidemic as reported by UNAIDS/WHO in May 2006, and refer to the end of 2007. (Data Source: *http://www.avert.org/worldstats.htm*)

Categories	Estimate	Range
People living with HIV/AIDS in 2007	33.2 million	30.6–36.1 million
Adults living with HIV/AIDS in 2007	30.8 million	28.2–33.6 million
Women living with HIV/AIDS in 2007	15.4 million	13.9–16.6 million
Children living with HIV/AIDS in 2007	2.5 million	2.2–2.6 million
People newly infected with HIV in 2007	2.5 million	1.8–4.1 million
AIDS deaths in 2007	**2.1 million**	**1.9–2.4 million**

Table 7-2 Regional statistics for HIV and AIDS end of 2007. (Data Source: *http://www.avert.org/worldstats.htm*)

Global Total	33.2 million children living with HIV/AIDS	2.5 million children newly infected	2.1 million adults and children
Sub-Saharan Africa	22.5 million	1.7 million	1.6 million
North Africa and Middle East	380,000	35,000	25,000
South, South-East, & Eastern Asia	4.8 million	432,000	302,000
Oceania	75,000	14,000	1,200
Latin America	1.6 million	100,000	58,000
Caribbean	230,000	17,000	11,000
Eastern Europe & Central Asia	1.6 million	150,000	55,000
Western and & Europe	760,000	31,000	12,000
North America	1.3 million	46,000	21,000

The epidemic continues to grow in this region, with over 1.6 million new infections between 2001 and 2007.

Since 1981, over **25 million** people have lost their lives to AIDS. The increasing number of women and children becoming infected with HIV is of great concern to epidemiologists. Young people, ages 15–24 years old account for half of all new HIV infections worldwide. It is estimated that more than 6,000 become infected with HIV every day.

By December 2007, women accounted for 50% of all adults living with HIV worldwide, and 61% in sub-Saharan Africa. Despite recent improvements in access to anti-retroviral treatment, 2007 saw approximately 2.1 million deaths from AIDS.

HIV/AIDS in Sub-Saharan Africa

The region of the world that is most affected by HIV and AIDS is sub-Saharan Africa. Nearly two-thirds of the world's HIV-positive people live in sub-Saharan Africa, although this region contains little more than 10% of the world's population. An estimated 22.5 million people in this region were living with HIV at the end of 2007 and approximately 1.7 million new infections occurred during that year. While this is a large number, it is 1 million less than the estimate at the end of 2005. Heterosexual sex is the main route of HIV transmission here, with women and girls accounting for

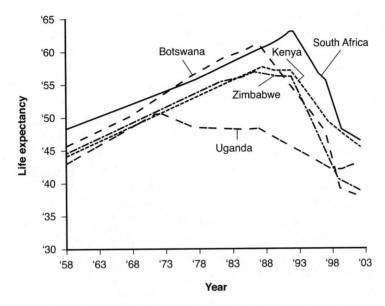

Figure 7-1. Decreased life expectancy due to AIDS epidemic in some hard-hit African countries. (Source: *http://en.wikipedia.org/wiki/AIDS*)

over 60% of all HIV infections. Most women with HIV in this region have been infected by their husbands or intimate partners.

In just the past year, the AIDS epidemic in Africa has claimed the lives of an estimated 1.6 million people in this region; and more than 11 million children have been orphaned by AIDS. Sadly, the extent of the AIDS crisis is becoming increasingly clear in many African countries, as the number of HIV-infected people progressing to AIDS begins to rise. It has been estimated that since the beginning of the epidemic more than 15 million Africans have died from AIDS.

In South Africa, Zambia, and Zimbabwe, young women (aged 15–24 years) are three to six times more likely to be infected than young men. This has significantly reduced life expectancy in many countries of the region (Figure 7-1). The epidemic claimed the lives of an estimated 2 million people in sub-Saharan Africa in just the past year. More than one in five of all pregnant women in six African countries (Botswana, Lesotho, Namibia, South Africa, Swaziland, and Zimbabwe) have HIV/AIDS. In Swaziland, nearly 40% of pregnant women are HIV positive. In sub-Saharan Africa, the estimated number of children under 18 orphaned by AIDS more than doubled between 2000 and 2007, currently reaching 12.1 million. Average life expectancy in sub-Saharan Africa is now 47 years; without AIDS, it would have been 62.

As an example, life expectancy at birth in Botswana has dropped to a level not seen since before 1950 and is expected to be only 26.7 by 2010. In less than 10 years time, many countries in southern Africa will see life expectancies fall to near 30—levels not seen since the end of the 19th century. At the end of 2007, there were 2.5 million children living with HIV around the world. It is estimated that there are 15 million children under 18 years old worldwide who have been orphaned by HIV/AIDS. Over 12 million of these children live in sub-Saharan Africa. At least a quarter of newborns infected with HIV die before age one, and up to 60% will die before reaching their second birthdays. At the end of 2007 the Joint United Nations Program on HIV/AIDS (UNAIDS) estimated that out of the 30.8 million adults worldwide living with HIV, around half are women, and in sub-Saharan Africa, women constitute 61% of people living with HIV. Gender inequalities as well as biological factors make women and girls especially vulnerable to HIV and to the impact of AIDS.

AIDS is now the leading cause of death in sub-Saharan Africa. Since the beginning of the AIDS epidemic 27 years ago, more than 20 million Africans have died from AIDS. There is a significant risk that some countries will be locked in a vicious cycle, as the number of people falling ill and subsequently dying from AIDS has a tremendous impact on many parts of African society, including demographic, household, health sector, educational, workplace, and economic aspects. Many grandparents who have lost all of their adult children to the disease are left raising their grandchildren, many of whom also are HIV positive. Clearly, this is a region that is crying out for help from the world community and if progress is not made quickly, whole generations will be wiped out by this devastating disease.

AIDS in Eastern Europe and Central Asia

The HIV epidemic in Eastern Europe and central Asia continue to grow, with 150,000 people newly infected with HIV in 2007, bringing to about 1.6 million the number of people in the region living with HIV. This is a twenty-fold increase in less than a decade. The death toll due to AIDS is also rising sharply; AIDS killed an estimated 55,000 adults and children in this region in 2007. The overwhelming majority of people living with HIV in this region are young; 75% of the reported infections between the years 2000 and 2004 were in people younger than 30 years old, while in Western Europe, the corresponding figure was 33% over the same period. The bulk of the people living with HIV in this region can be found in two countries: the Russian Federation and Ukraine. The epidemic continues to grow in the Ukraine, with significantly more new HIV diagnoses occurring each year, while the Russian Federation has the largest region of the AIDS epidemic in all of Europe, constituting massive prevention, treatment, and care challenges. At the end of 2007 there were approximately

940,000 people living with HIV in the Russian Federation. Several central Asian republics are experiencing the early stages of the AIDS epidemic, while increasing levels of risky behavior in southeastern Europe suggests that HIV could strengthen its presence there unless prevention efforts are stepped up. Discrimination against vulnerable groups is evident in the Russian Federation, where more than 90% of the people living with HIV were infected through injection drug use, but represent only 13% of those receiving antiretroviral therapy.

HIV/AIDS in Oceania

Oceania collectively encompasses all of Australia, New Zealand, Papua New Guinea, as well as the thousands of coral atolls and volcanic islands of the South Pacific Ocean, including the *Melanesia* and *Polynesia* groups and all the islands of Micronesia. In Oceania there were an estimated 75,000 people living with HIV at the end of 2007. Over this period 14,000 people became newly infected with HIV and an estimated 1,200 died of AIDS-related illnesses. The epidemic in **Papua New Guinea** accounts for more than 70% of all HIV infections reported to date in Oceania, with an adult infection rate of 1.3%, and the epidemic is growing at an alarming rate. HIV diagnoses in this region have been increasing by approximately 30% annually since 1997. The majority of reported HIV infections to date have been in rural areas, where more than 80% of the population lives. Unsafe heterosexual intercourse is estimated to be the main mode of HIV transmission. Australia and New Zealand have relatively small HIV epidemics with approximately 0.2% of their populations affected. In **Australia** an estimated 16,000 adults and children were living with HIV in 2005. After declining in the late 1990s, annual new HIV diagnoses are again approaching earlier levels. Unprotected sex between MSM accounts for the greatest increase in new cases of HIV infections reflecting a revival of high-risk sexual behavior.

Annually, new diagnoses in **New Zealand** have more than doubled since 1999, when they had fewer than 80, to 183 in 2005. The national adult HIV prevalence, however, remains very low at under 0.2%. Much of the recent trend is attributable to an increase in HIV diagnoses among MSM. HIV-infection levels are low in the rest of Oceania, but this could change, especially on the islands of Polynesia. On **Vanuatu**, more than 40% of pregnant women have been found to have at least one sexually transmitted infection, as did 43% of pregnant women in **Samoa**'s capital, Apia. In Dili, **Timor-Leste**, 60% of sex workers tested positive for at least one sexually transmitted infection, as did almost 30% of taxi drivers and MSM (*http://w3.unaids.org/en/Regions_Countries/default.asp*).

HIV/AIDS in Asia

Overall in Asia, an estimated 4.9 million [3.7 million–6.7 million] people were living with HIV in 2007. This number includes the 440,000 [210,000–1.0 million] people who became newly infected in the past year. Approximately 300,000 [250,000–470,000] died from AIDS-related illnesses in 2007. These latest estimates are significantly lower than those reported in 2005, when approximately 8.3 million people were estimated to be living with the disease. This dramatic change in numbers is due mainly to the revised estimates from **China** and more importantly from **India,** where it was estimated that more than two-thirds of Asia's HIV infected patients reside. New, more accurate estimates of HIV indicate that approximately 2.5 million (2 million–3.1 million) people in **India** were living with HIV in 2006, with national adult HIV prevalence of 0.36%. Although national HIV infection levels in Asia are low compared with those of some other continents, notably Africa, the populations of many Asian nations are so large that even low national HIV prevalence means large numbers of people are living with HIV. The 92,000 [21,000–220,000] adults and children estimated to be newly infected with HIV in **East Asia** in 2007 represent an increase of almost 20% over the 77,000 [49,000–130,000] people who acquired HIV in 2001. Like many other regions in the world, risky behavior seems to be the sustaining power to this growing epidemic. In Asia, there is often more than one form, but at the very heart lies the interplay between injection-drug use and unprotected sex, especially commercial. Cambodia and Thailand has seen an upsurge in HIV infection in MSM, and injection-drug use, while countries like Indonesia and Pakistan will certainly follow along with similar statistics if they do not urgently scale up their responses. In **Thailand,** national adult HIV prevalence was estimated at 1.4% in 2005. Prevention efforts have resulted in declining levels of HIV in this country since the late 1990s. This decrease was mainly due to fewer men buying sex and the steady increase in condom use. However, recent studies show that premarital sex has become more commonplace among young Thais and more than one-third of HIV infections in 2006 were among women who had been infected by their long-term partners. Despite the overall achievements in reversing the HIV epidemic in Thailand, prevalence among injecting drug users has remained high over the past 15 years, ranging between 30% and 50% (WHO, 2007).

There are currently an estimated 700,000 people living with HIV in China, including about 75,000 AIDS patients. Injection-drug users account for almost half (44%) of people living there with HIV. Almost one-half of China's injection-drug users share needles and syringes, and one in ten also engage in high-risk sexual behavior. While unofficial estimates out the number at three to four million, in 2007 there were a reported 937,000 registered injecting drug users in China. As HIV spreads from drug users, sex workers, and their clients to **Asia's** general population, the proportion of HIV infections in women has also been increasing. In 2007, for the first time ever, unsafe sex overtook drug use as the main cause of new HIV infections in the country. This confirms fears that the epidemic is moving outward from the high prevalence groups traditionally associated with the HIV epidemic in China. **China's "Four Frees and One Care" program,** which offers free HIV testing, free antiviral drugs to rural residents with AIDS, free drugs to pregnant women, free schooling for AIDS orphans, care and economic assistance to affected households, may provide a model for other nations in supporting families and societies affected by AIDS. UNAIDS estimates that in 2006, around 25% of those in need of antiretroviral drugs in **China** were receiving them.

HIV/AIDS in the Caribbean

The region covering the Islands of the Caribbean is the area second-most affected by AIDS in the world. Adult HIV prevalence in the Caribbean was estimated at 1.0% [0.9%–1.2%] at the end of 2007. Prevalence in this region is highest in the Dominican Republic and Haiti, which together account for nearly three-quarters of the 230,000 [210,000–270,000] people living with HIV in the Caribbean. It is estimated that there were 17,000 [15,000–23,000] new HIV infections in 2007. An estimated 11,000 [9,800–18,000] people in the Caribbean died of AIDS in 2007 and AIDS remains one of the leading causes of death among persons aged 25 to 44 years. The islands of the Bahamas, along with Belize, Guyana, Haiti, and Trinidad and Tobago seem to be hardest hit by AIDS. Most cases in this region are due to heterosexual contact, with the exception of Puerto Rico and Bermuda, where injection-drug use accounts for a large share of the epidemic. The epidemic is most pronounced in Haiti, where condom use is very low despite knowledge of HIV/AIDS. In the Bahamas and Barbados, there are signs that stronger prevention efforts are beginning to help lower HIV infection rates. The HIV epidemics in Jamaica, the Bahamas, and Trinidad and Tobago have also been stable over recent years. The Caribbean's status as the second-most affected region in the world shrouds substantial differences in the islands that are affected. Estimated national adult HIV prevalence is about 1% in Barbados, Dominican Republic, Jamaica, and Suriname; 2% in the Bahamas, Guyana, and Trinidad and Tobago; and exceeds 3% in Haiti. In Cuba, on the other hand, prevalence has not reached 0.2%. The region's epidemics seem to be driven primarily by unprotected heterosexual intercourse, which is the documented mode of transmission in over 75% of all AIDS cases reported to date. Commercial sex is a prominent factor in some islands, against a backdrop of severe poverty, high unemployment, and gender inequality issues. Only a small minority of infections is currently linked to injection-drug use and these cases are mostly in Bermuda and Puerto Rico. While universal treatment access is being achieved in Cuba, and coverage is relatively high in the Bahamas and Barbados, access to treatment is poor in the most affected islands in the Caribbean.

HIV Epidemic in Latin America

The HIV epidemics in Latin America remain generally stable, and HIV transmission continues to occur among populations at higher risk of exposure, including sex workers and men who have sex with men. There were a total of 1.6 million people living with HIV in Latin America in 2007—more than in the United States, Canada, Japan, and the United Kingdom combined and some 100,000 people became newly infected with the virus. Around 39,000 children under the age of 15 were living with HIV in 2007 and AIDS claimed some 58,000 lives. Women accounted for 510,000 of those living with HIV at the end of 2007 in Latin America. The growth of the epidemic in this region seems to stem from increased rates in Brazil, which accounts for more than one-third of the people living with HIV in Latin America. The serious growth in Brazil stems from MSM, injection-drug users, and people who have heterosexual sex. The light at the end of the tunnel in Brazil is that the country strives to help its people living with HIV; everyone living with HIV has access to antiretroviral drugs through the national health system. The Brazilian government estimates that antiretroviral treatment has contributed to a 50% fall in mortality rates, a 60–80% decrease in morbidity rates, and a 70% reduction in hospitalizations among HIV-positive people. In Argentina and Uruguay, HIV remains largely an urban problem. The HIV epidemic is also growing in Venezuela, where the virus is spreading mainly through unsafe heterosexual sex. In Bolivia, infections are mainly among sex workers and MSM. Like many other regions of the world, however, HIV/AIDS in Latin America remains a significant problem. The epidemic in this region is sometimes referred to as a *"hidden"* crisis. Awareness remains low, governments have been relatively inactive, surveillance of those affected is sometimes unreliable and the associated stigma has stopped many from conducting open and frank debate about this problem. There is no doubt that prevention strategies need to be put in place along with mechanisms to tackle discrimination. The good news is that there has been tremendous success in providing anti-retroviral drug treatments for those affected in the region.

HIV/AIDS in North America and Western Europe

In North America and Western Europe, the number of people living with HIV has continued to grow, although the number of AIDS deaths remains comparatively low. This is in part because of the availability of antiretroviral therapy coupled with the fact that many of those infected can afford to pay for these drugs. At the end of 2007 there were 2.1 million people living with HIV in this region accounting for 6.4% of the world's AIDS cases. There were approximately 78,000 new cases of HIV infections and 32,000 AIDS deaths in 2007.

Reports indicate that more than half a million people are living with HIV in Western Europe. This number seems to be growing with the resurgence of risky sexual behavior in several countries of the region. Heterosexual contact has emerged as the dominant cause of new HIV infections in several countries in Western Europe, accounting for 42% of new HIV infections diagnosed in Western Europe in 2006. A substantial proportion of new diagnoses are in immigrants originating from countries with serious epidemics, principally countries in sub-Saharan Africa. As a result of anti-retroviral drug availability, the number of AIDS deaths plummeted in the late 1990s. In Western and Central Europe, the **United Kingdom** continues to have a large HIV epidemic, together with **France, Italy,** and **Spain.** In other countries in this region, approximately one-third of persons (32% in 2005) newly diagnosed with HIV are unaware of their infection (Health Protection Agency, 2006).

HIV incidence peaked in the United States in 1993 and then started a steady decline. The annual numbers of AIDS diagnoses have been relatively constant since 2000, with an estimated 37,852 in 2006. In total, an estimated 1,014,797 people have been diagnosed with AIDS in America. The estimated number of people living with HIV in the United States at the end of 2003 exceeded one million for the first time, while in Canada, just under 58,000 HIV diagnoses were reported at the end of 2004. The death rate among people with AIDS has also remained relatively stable in recent years; there were an estimated 14,627 deaths in 2006. Since the beginning of the epidemic, an estimated 565,927 people with AIDS have died in the United States.

HIV/AIDS in African Americans

Currently the epidemic is disproportionately spreading among African Americans compared with members of other races, and is presently a major health crisis. The epidemic has also had a disproportionate impact on black women, youth, and men who have sex with men, and its impact varies across the country. Today, there are approximately 1.2 million people living with HIV/AIDS in the United States, including more than 500,000 who are Black. Analysis of national household survey data found that more than 2% of Blacks in the United States were HIV positive, higher than any other group. Data from the 2000 census indicate that Blacks make up approximately 12% of the U.S. population. However, in 2006, Blacks accounted for 49% of the estimated all new cases of HIV/AIDS diagnoses in the United States in the 33 states with long-term, confidential name-based HIV reporting (Figure 7-2). The primary transmission category for black men living with HIV/AIDS was sexual contact with other men, followed by injection-drug use and high-risk heterosexual contact. For black women living with HIV/AIDS, the primary transmission category is high-risk heterosexual contact, followed by injection-drug use.

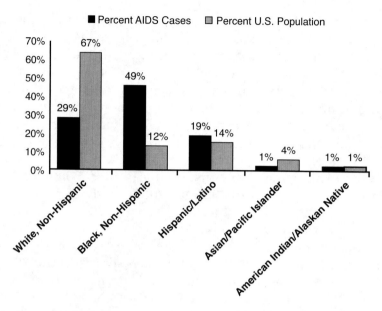

Figure 7-2 Estimated AIDS diagnoses and U.S. population by race/ethnicity. Note that African Americans who make up 12% of the U.S. population account for half of all new HIV infections in 2006. (Source: This figure was generated from data distributed by the Keiser Family foundation)

This rate of AIDS diagnosis among Backs has risen significantly from 25% of all cases in 1985 to 49% in 2006. The number of African Americans living with AIDS increased by 27% between 2002 and 2006, compared to a 19% increase among their white counterparts. In addition, 65% of all infants infected with HIV perinatally in 2006 were black. African American women account for the majority of new AIDS cases among women (66% in 2006); white and Latina women accounted for 17% and 16% of new AIDS cases, respectively. HIV/AIDS was the fourth leading cause of death for Black men and third for Black women, aged 25–44, in 2004, ranking higher than any other racial or ethnic group in the United States. African American teens aged 13–19 accounted for 69% of new AIDS cases reported in 2005, although they represent only 16% of U.S. teens. To further complicate an already serious situation, African Americans with HIV/AIDS may face greater barriers when trying to access health care.

The AIDS case rate per 100,000 among African American adults and adolescents was more than 9 times that of Whites in 2006 (Figure 7-3). The AIDS case rate for black men (82.9) was the highest of any group, followed by black women (40.4). By comparison, the rate among white men was 11.2. While both white and black American males are more likely to have been infected through sex with other men, heterosexual transmission and injection drug use account for a greater share of infections among

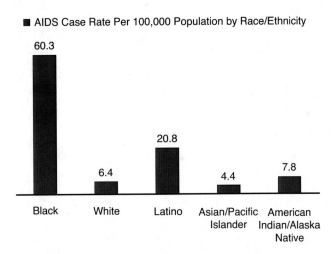

■ AIDS Case Rate Per 100,000 Population by Race/Ethnicity

Figure 7-3 AIDS case rate per 100,000 of the population, by race/ethnicity, for U.S. adults/adolescents in 2006.

black men than white men. A recent study in five major U.S. cities found that 46% of black MSM in the study were infected with HIV, compared to 21% of white MSM and 17% of Latino MSM. This disturbing trend of HIV infection among African Americans in general appears to be influenced by several factors. These include but are not limited to: (1) access to and use of health care system; (2) lack of health insurance; and (3) geography [10 states account for 71% of all HIV cases among blacks]. It is clear that something needs to be done immediately to address this growing epidemic among African Americans. If this is not done very soon, the HIV/AIDS epidemic will again become a very serious health care crisis in the United States while everyone is donating resources to and focusing on the problem in Africa and Asia.

CHAPTER 8

Testing and Diagnosis of HIV/AIDS

The first HIV tests developed in 1985 were designed to screen blood for transfusion purposes. Back then, there were no treatments available for HIV infection and it was not even known if an HIV-infected person would get AIDS or how quickly it might happen. It seemed at the time that there was no pressing reason for people to find out their HIV status. There were people, however, who, based on their activities, suspected that they might be infected and wanted to know their status. It was feared that these individuals were going to donate blood just to get tested. That soon changed with the advent of antiretroviral therapy and prophylaxis (preventive treatment) aimed at preventing opportunistic infections. Over time, HIV testing became a gateway to treatment, as well as a prevention tool. There are specific tests available that can readily determine whether or not you are infected with HIV.

Although there have been many technological advances since 1985 where testing is concerned, HIV testing still follows the same basic testing procedure as in 1985: HIV infection is only considered confirmed after two tests have been done—a screening test and a confirmatory test. These tests look for "antibodies" to HIV. These antibodies are proteins produced by the immune system to fight an infection. Most antibody tests are used for diagnostic purposes to determine if someone is infected. There are, however, additional HIV tests that are used to monitor the status of a person who already knows that he or she is infected with HIV. These help measure how quickly the virus is multiplying (viral load test) or the health of your immune system (CD4 cell count). HIV testing is integral to prevention, treatment, and care efforts. Knowledge of one's HIV status is important for preventing the spread of disease. Test results can also reduce stress and undue anxiety if your result is negative. If you are HIV positive, testing provides an opportunity to receive risk reduction counseling and other useful information. Many who learn they are HIV positive modify their behavior to reduce the risk of HIV transmission. Early knowledge of HIV status is also important for linking those who are HIV positive to medical care and services that can reduce morbidity and mortality and improve quality of life. Individuals infected with HIV normally develop detectable antibodies within 3 months after infection; however, it can take longer depending on the person's immune system and general health status.

HIV Testing for Pregnant Women

Approximately 25% of HIV-infected pregnant women who are not treated during pregnancy can transmit HIV to their infants during pregnancy, during labor and delivery, or through breast-feeding. In 1998, the Institute of Medicine (IOM) published a report that recommended simple, routine, and voluntary HIV testing for all pregnant women in antenatal settings, given the effective interventions available to treat HIV-infected women and reduce risk for perinatal HIV transmission. This HIV test should be offered to all pregnant women as part of the standard battery of prenatal tests, regardless of risk factors and the prevalence rates in the community. The IOM report clearly stated that whenever a woman of childbearing age is unaware of her HIV status or her risk for HIV or when an HIV-infected pregnant woman does not know her status, an opportunity is missed. This means that the woman (a) does not receive prenatal care, (b) is not offered HIV testing, (c) is unable to obtain HIV testing, (d) is not offered chemoprophylaxis, (e) is unable to obtain chemoprophylaxis, or (f) does not complete the chemoprophylaxis regimen. Prophylaxis failures occur when an infant becomes infected despite chemoprophylaxis and other preventive interventions. The report concluded that each of these missed opportunities or failures deserves attention from the appropriate service providers and available prevention programs (*http://www. cdc.gov/mmwr/preview/mmwrhtml/rr5019a2.htm*).

Screening and Confirmatory Tests for HIV

The most common screening test available for HIV infection is the **enzyme-linked immunosorbent assay (ELISA),** sometimes called enzyme immunoassay (EIA). The most often used confirmatory test is the **Western blot.** These tests are by no means unique or limited to HIV testing. Identical technology is used in tests for numerous illnesses, including Lyme disease, hepatitis, and cytomegalovirus. The body does not naturally have HIV antibodies. Antibodies are only produced in the presence of an infection. Hence, if HIV antibodies do exist, it is a sign that HIV has invaded the body. Both the ELISA test and the Western blot test detect HIV infection by detecting the presence of HIV antibodies. Both of these tests rely on the fact that HIV and HIV antibodies bind together. The binding site on the surface of HIV is a protein antigen. There is an area on the surface of HIV antibodies into which HIV antigens fit, like a key fits into a lock.

The ELISA technique is used for initial screening because it possesses high sensitivity, uses a small volume, is high throughput, and is fully automated, which means there is less handling of the samples. The ELISA test uses recombinant (artificial) HIV proteins immobilized on a solid surface (typically inside a plastic 96 well plate) that are able to capture antibodies to the virus contained in the infected blood of the patient. Captured antibodies can be detected using a second antibody that is specific for any captured antibodies bound to HIV proteins. Since there are wash steps between each addition of patient serum or reagent, only antibodies bound to proteins on the plate surface will be present. The second antibody is usually coupled to an enzyme (hence, the name) that when introduced to its specific substrate (molecules upon which an enzyme acts) produces a colored product (Figure 8-1). The change in color

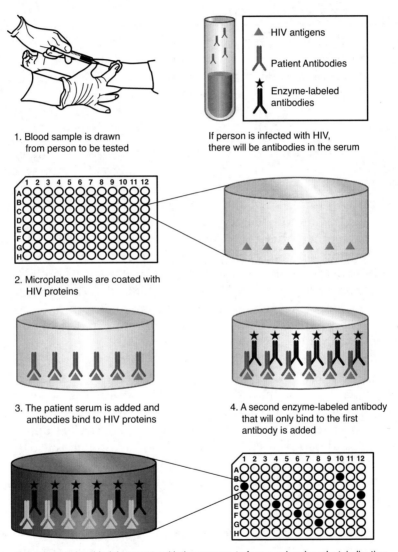

1. Blood sample is drawn from person to be tested

▲ HIV antigens

Patient Antibodies

★ Enzyme-labeled antibodies

If person is infected with HIV, there will be antibodies in the serum

2. Microplate wells are coated with HIV proteins

3. The patient serum is added and antibodies bind to HIV proteins

4. A second enzyme-labeled antibody that will only bind to the first antibody is added

5. A substrate is added that reacts with the enzyme, to form a colored product, indicating that patient antibodies are present.

Figure 8-1 Diagram of an ELISA to detect HIV-specific antibodies. The intensity of the color in each well is directly proportional to the amount of HIV antibodies in the patient's blood.

Table 8-1 HIV proteins typically seen reacting with patient antibodies on a Western blot. Please note that there might be more proteins than these on any positive blot. Gene products from each area of the virus are represented here (envelope, capsid, and core proteins).

Protein	Origin and Function
gp160	viral envelope precursor (env)
gp120	viral envelope protein (env) binds to CD4
p24	viral core protein (gag)
p31	Reverse Transcriptase (pol)

is read by a machine that determines the optical density based on the intensity of the color. The intensity of this color is directly proportional to the amount of HIV-specific antibodies in the patient's blood. This test can be performed on either blood or urine. If the ELISA test is positive, then the confirmatory Western blot test is done. The Western blot is also an antibody-based test; however, this test is more specific than the ELISA. HIV, like any other virus, is composed of a number of different proteins. HIV positivity can therefore only be confirmed by the presence of these proteins. If HIV antibodies to proteins from all parts of the virus are present in the patient's sample, this confirms seropositivity and there is very little doubt that the individual is infected with the virus. The antibodies seen in a positive sample are typically reactive with all classes of HIV gene products. Table 8-1 lists the proteins most likely to react with patient antibodies from infected sera on a Western blot.

For the Western blot, the recombinant HIV proteins are separated by electrophoresis on a sodium dodecyl sulfate polyacrylamide gel electrophoresis (SDS-PAGE) system based on their individual molecular weight. The larger proteins are at the top of the gel matrix and the smaller proteins that travel faster are toward the bottom. Once these proteins have been separated on the gel, they are transferred by electrical charge to a solid paper-based membrane. This membrane is usually made from nitrocellulose or polyvinylidene fluoride (PVDF). The proteins will be transferred in the same pattern as seen on the gel. The membrane is then cut into strips to facilitate testing of a large number of samples for antibodies directed against the blotted protein (antigen). The patient's serum sample is then taken and added to the paper membrane with the HIV proteins on it and allowed to interact. If the patient is HIV positive, there will be antibodies made against various parts of the virus floating around in the serum. Each antibody will react with the specific protein against which it was made on the blot membrane. Since these proteins (antigen & antibodies) are so small, you will not be able to see this reaction with the naked eye or even with a microscope; therefore, some type of "label" must be placed on this reacting complex to visualize the reaction.

When a labeled secondary antibody is added in a similar manner as the ELISA test above, it will bind to the patient antibodies that are bound to the membrane. A substrate will then be added and a colored product will be formed, resulting in a dark precipitate at the sites where there are bound antibodies on the blot.

Interpretation of Western Blot Data

In 1997 the Centers for Disease Control along with several other organizations established criteria for serologic interpretation of HIV Western blot tests (See Figure 8-2). While different countries have their own standards for what constitutes a positive HIV Western blot, the CDC and FDA have arrived at a decision that must include a specific combination of HIV proteins. If no viral bands are detected, the result is negative (Table 8-2). If at least one viral band for each of the gag, pol, and env gene-product groups is present, the result is positive (See Table 8-1). Western blot tests in which less than the required number of viral bands are detected will be reported as indeterminate. A person who has an indeterminate result should be retested, since later tests may be more conclusive.

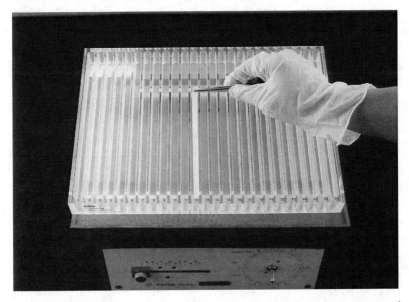

Figure 8-2 A technician is placing a Western blot strip in a test tray containing patient serum. Antigens have already been transferred from the gel onto the test strip. Antibodies, if present in the patient serum, will bind to this sheet and later be detected with a labeled secondary antibody. (Source: CDC, *http://phil.cdc.gov/ phil_images/20021205/23/PHIL_2613.tif*)

Table 8-2 Table of banding patterns on a Western blot and interpretations that would be reported.

Banding Patterns Observed	Reported Results
No bands present	Negative
Bands at either p31 OR p24 AND bands present at either gp160 OR gp120	Positive
Bands present, but pattern does not meet criteria for positivity	Indeterminate

Research data show that almost all HIV-infected persons with indeterminate Western blot results will develop a positive result when tested one month after the first test. If an individual remains persistently indeterminate over a period of six months, this would suggest that the results are not due to HIV infection and therefore are the result of a false-positive reaction. Indeterminate Western blot results can be caused by either incomplete antibody response to HIV in samples from infected persons or non-specific reactions in samples from uninfected persons.

Incomplete antibody responses that produce negative or indeterminate results on Western blot tests can occur among persons recently infected with HIV who have low levels of detectable antibodies (i.e., seroconversion), persons who have end-stage HIV disease, and perinatally exposed but uninfected infants who are seroreverting (i.e., losing maternal antibody). Nonspecific reactions producing indeterminate results in uninfected persons seem to occur more frequently among pregnant or parous women than among other persons. False-positive Western blot results (especially those with a majority of bands) are rare.

False positive tests results using the ELISA tests will require secondary testing using this Western blot, immunoblotting procedure. The bound antigens are detected when tagged with antibodies during analysis of the nitrocellulose sheet. Western blot results help confirm if someone is HIV positive because some conditions (including lupus and Lyme disease) may yield a false-positive ELISA test result. The blots of real patient data along with positive and negative controls shown in Figure 8-3 are very typical of what is seen in the clinical setting. While it is clear that sample number 1 is negative and that number 3 is a clear positive, sample 2 does not have either a gp120 or a gp160 band to represent the envelope proteins. It therefore is an indeterminate result and the patient in this case would be asked to repeat the test in approximately one month. An HIV test should be considered positive only after screening (ELISA) and confirmatory tests (Western blot) are reactive. A confirmed positive test result indicates that a person has been infected with HIV. Please note, however, that it does not mean that the person has AIDS. Incorrect HIV test results occur primarily because of specimen-handling errors, laboratory errors, or failure to follow the recommended testing algorithm.

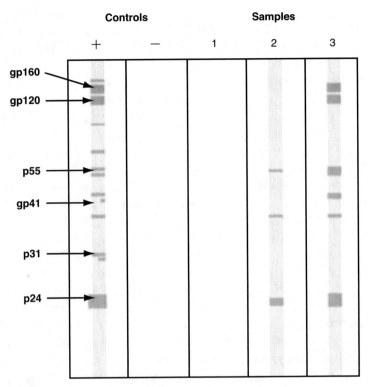

Figure 8-3 Picture of HIV-specific Western blot showing controls and patient samples. Sample number 1 is negative, while sample number 3 is positive based on the CDC criteria. Sample 2 falls into the category of indeterminate result and would be retested before the results can be confirmed.

Nucleic Acid-Based HIV Testing

Both the ELISA and Western blot tests are indirect assays, meaning they do not look for HIV. Instead, they detect the immune system response to HIV; in the absence of a developed immune system, they would be ineffective. This would be the case if the HIV status of an infant needed to be determined. Furthermore, neither the Western blot nor the ELISA protocols help the physician determine the viral load. As discussed in the previous section dealing with the progression to AIDS, determination of the level of HIV in the peripheral blood is very important in establishing disease stage and treatment options, as well as the effectiveness of a treatment regimen.

The most effective tests that directly assay for the presence of the HIV particles involve the use of the very versatile **polymerase chain reaction (PCR)**. The PCR chemically multiplies viral DNA that exists in the sample by a factor of approximately one million (this is the "chain reaction," which its name describes). The PCR reaction can detect HIV DNA or RNA in the patient's blood. The test is designed to detect and amplify a specific fragment of the HIV DNA in a relatively short period of time (Figure 8-4).

Since the viral RNA is first converted into double-stranded DNA before being integrated into the host cell DNA or genome, PCR may be used to detect this proviral DNA. For this type of test, blood is drawn from the patient's arm; the peripheral blood mononuclear cells (PBMCs), which are circulating white blood cells, are isolated. DNA is then isolated from the cells and PCR amplification of a specific short fragment is accomplished using a machine called a Thermocycler. This amplified DNA is then quantified to determine the viral load. There are PCR protocols available that can detect 1 cell containing proviral DNA in a sample of 150,000 cells. HIV infection can be reasonably excluded in an infant if PCR tests performed at one and four months are negative for HIV DNA. Definitive exclusion of infection does not take place, however, until HIV antibody testing is negative. Infants are tested for persistent HIV antibody at about 12 months of age, but passive maternal antibody may be present until 18 months since mom's antibodies can cross the placenta and go into the baby's circulation to protect it or may come from breast milk. Detection of HIV antibody in a child after 18 months would be diagnostic of HIV infection.

In 1995, a series of papers were published on the natural history of HIV infection, which revealed that there is rapid replication and turnover of HIV, even in the clinically "latent" phase. As a result, **tests to measure viral RNA levels** were developed and became useful tools in monitoring the response of HIV to therapy in adults. These same tests also became attractive tools in the diagnosis of vertically transmitted infection. In this test, RNA is isolated from the blood sample and reverse transcription PCR (RT-PCR) is used to convert viral RNA into DNA. PCR is then used as described for the proviral DNA to amplify this DNA. This test can quantify the amount of viral RNA in the blood and can detect as little as 50 copies of viral RNA per milliliter of blood, which translates to 25 viral particles since each viral particle contains two copies of RNA. New tests continue to be developed each day and recently a Real-Time Immuno-PCR test, which combines parts of traditional antibody testing with PCR, has been developed. This new test can reportedly detect the virus when only two copies of HIV are present per milliliter of blood (*http://www.hivandhepatitis.com/recent/test/real-time/061604f.html*). This would allow clinicians to detect HIV in the blood 12 days after a person has been exposed to the virus!

Rapid HIV Tests

The major drawback of the HIV tests discussed above is that they take a long time before the results become available to the physician and the patient. Most test results

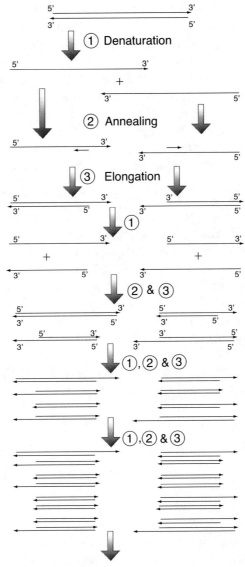

Exponential growth of short product

Figure 8-4 Diagram of PCR reaction to demonstrate how amplification leads to the exponential growth of a short product flanked by the primers. (1) Denaturing at 94–96°C; (2) Annealing at ~65°C; (3) Elongation at 72°C. Four cycles are shown here. The longer lines represent the DNA template to which primers (shorter arrows) anneal that are extended by the DNA polymerase (light circles), to give shorter DNA products (directional lines), which themselves are used as templates as PCR progresses.

become available anywhere from one to two weeks after the blood is drawn. The FDA recently approved a number of rapid HIV tests, which function on the same principle as the ELISA, but the results are available in less than one hour. Some of the major advantages of the rapid tests are increased numbers of people benefit from knowing their HIV status; there is an increased uptake of results by people being tested; and less reliance is placed on laboratory services for obtaining the results.

In March 2004, the FDA approved the **OraQuick** Rapid HIV-1 Antibody Test for use with oral fluid and plasma specimens. The test was previously approved for use with blood samples in 2002. Although there are several rapid HIV tests on the market, this is the only rapid HIV test to be approved in the United States by the FDA for use with oral fluid. The OraQuick Advance test can now be used with oral fluid specimens taken from the mouth, with plasma or with whole blood. The test is performed by using the device to gently and completely swab both upper and lower outer gums one time around. The device is then inserted into a vial containing a developer solution. After 20 minutes, the test device is visually examined for the results. If HIV antibodies are present in the solution, two reddish-purple lines will be displayed in a small window in the device. Rapid tests are single-use, and do not require laboratory facilities or highly trained staff. This makes rapid tests very suitable for use in resource-limited countries. It also allows testing, counseling, and referrals to be done in one visit. This test is not yet available for home use. As is true of current ELISA antibody procedures, an initial reactive rapid HIV test result should be confirmed by Western blot. The patient should be in a position to receive the necessary counseling when the results are presented and follow-up visits and care should be arranged.

Tests for HIV Progression

Various tests are available to monitor HIV disease progression and the overall health of the infected person. Tests for HIV viral load can provide a good picture of viral activity, and relative viral quantity in the circulation, while CD4 cell counts monitor the status of the immune system and may assist physicians in predicting and subsequently preventing the onset of opportunistic illnesses. A viral load test in conjunction with a CD4 cell count can help guide treatment decisions and indicate whether treatment is working.

Viral load describes the amount of HIV particles in the circulating blood and is expressed either as copies of RNA per milliliter of blood (copies/mL) or as logs. A viral load of 100,000 copies/mL or greater is considered high, while levels below 10,000 copies/mL are considered low. Research has consistently shown that higher viral loads are associated with more rapid HIV disease progression and an increased risk of death. Current U.S. HIV treatment guidelines recommend that people should consider starting antiretroviral treatment if their viral load is above 55,000 copies/mL. Sometimes with successful treatment, the level of HIV may be too low to be measured, and the viral load is said to be undetectable, or below the limit of quantification. This, however, does not mean that HIV has been eradicated and the person is now disease-free. People with undetectable viral load maintain very low viral levels in circulating blood cells. Moreover, the circulating levels of HIV are much less than amounts found in immune organs such as the lymph nodes where there is a higher concentration of affected cells. In addition, even when HIV is not detectable in the blood, it may be detectable in the semen, female genital secretions, cerebrospinal fluid, and other tissues. Effective anti-HIV treatment can often reduce viral load to low or undetectable levels. Therapy that does not produce an undetectable viral load is often said to be failing, possibly because of viral mutation.

The **CD4 T-cell count** is not an HIV test, but rather a procedure where the number of CD4 T cells in one microliter of blood is counted in a standard medical lab test after a blood draw. This CD4 test does not check for the presence of HIV. It is used to monitor the immune system function in HIV-positive people by measuring the declining CD4 T cell. In HIV-positive people, AIDS is officially diagnosed when the count drops below 200 cells or when certain opportunistic infections occur. This use of a CD4 count as an AIDS criterion started in 1992. This 200 count is not arbitrary, but was chosen because it corresponds with an increased likelihood of opportunistic infections. Lower levels of CD4 counts in people with AIDS are indicators that prophylaxis against certain types of opportunistic infections should be instituted. Generally speaking, the lower the number of T cells, the lower the immune system's function will be. This means that the body will be more vulnerable to opportunistic infections. Normal CD4 T-cell counts are between 1,000 and 1,500 CD4+ T cells per microliter and the counts may fluctuate in healthy people, depending on recent infection status, nutrition, exercise, and other factors—even the time of day. Women tend to have somewhat lower counts than men.

CHAPTER 9

HIV/AIDS Treatment

By 1986, AIDS researchers could describe the nature and shape of HIV proteins as well as key molecules on the immune system cells that HIV infects. Researchers began to test drugs already approved to treat other conditions to see if any of them would have an effect on the virus. They soon discovered Azidothymidine (Zidovudine, see Figure 9-1), known as AZT, that was capable of abrogating the action of the reverse transcriptase enzyme used by HIV to change its RNA into DNA so that the provirus DNA may integrate into the host genome.

The blocking of reverse transcriptase significantly slowed the progression of the virus. Antiretroviral (ARV) therapy is the main type of treatment for HIV or AIDS. The goal of anti-HIV treatment is to reduce the amount of HIV in the body. It is not a cure; however, it can significantly slow the progression to AIDS and millions of people take these drugs each day. As of December 2006, an estimated 7.1 million of the people living with HIV, especially in low- and middle-income countries were urgently in need of life-saving ARV medication. Of these only one in four (2.0 million) were being treated with these drugs.

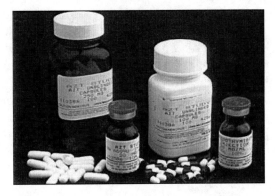

Figure 9-1 AZT (*zidovudine*), the first medication shown to be effective against HIV. (Source: NIH)

Nucleoside and Non-Nucleoside Inhibitors

There are two types of reverse transcriptase inhibitors—nucleoside and non-nucleoside—which target slightly different parts of the enzyme. **Nucleoside reverse transcriptase inhibitors (NRTIs)** such as zidovudine (AZT) are similar in structure to nucleosides, which are the building blocks of nucleic acid (analogs). These drugs differ slightly from nucleosides; hence, if they become incorporated into the growing DNA strand, they will interfere with the DNA replication process. Specifically, NRTIs incorporated into the growing DNA strand terminate further strand elongation. The enzyme responsible for human cellular DNA replication, DNA polymerase, is generally able to differentiate between the analogs and the real nucleotides, therefore avoiding incorporation into our DNA. The HIV reverse transcriptase is much less picky and routinely utilizes these analogs when converting viral RNA into DNA. Viral replication is therefore blocked.

Non-nucleoside reverse transcriptase inhibitors (NNRTIs), such as tenofovir DF (Viread) and stavudine (Zerit), are a chemically diverse class of drugs that bind a pocket near the active site of reverse transcriptase and as such inhibit the enzyme. The viral genetic material therefore cannot be incorporated into the genetic material of the cell, resulting in a lack of virus production. Since we do not normally have reverse transcriptase in our system, these drugs do not disrupt cellular functions. The action of NRTIs and NNRTIs on HIV was short lived. The virus soon came back to life as if the drug was not even present. The virus had mutated and the drug no longer had an effect. Even when these reverse transcriptase inhibitors were used together, they had no effect on the mutated form of the virus and the health of the patient soon deteriorated.

Researchers soon came up with the idea that a multipronged approach might be best since this had worked for other rapidly mutating microbes like tuberculosis. By 1996, a new drug target was found. They devised a way to inhibit the action of another HIV enzyme called protease. Protease inhibitors stop the virus from forming mature virions, or individual virus particles by blocking the action of the enzyme on the new proteins that are formed, making the virus particles nonfunctional. Protease inhibitors [lopinavir/ritonavir (Kaletra)] seem to work miracles when used in combination with two reverse transcriptase inhibitors.

Entry Inhibitors

All reverse transcriptase inhibitors and protease inhibitors act on HIV once it enters the cell. Everyone would agree that it would be better to prevent HIV entry in the first place and, consequently, there has been a concerted effort to develop entry inhibitors. Entry inhibitors work by attaching themselves to proteins on the surface of T cells or proteins on the surface of HIV, thus preventing entry into healthy cells of the body. In order for HIV particles to bind to T cells, the proteins protruding through the viral envelope must bind to the proteins on the surface of the T cells. Entry inhibitors would prevent this from happening. Some entry inhibitors target the gp120 or gp41 proteins on the surface HIV, while others target the CD4 protein, the CCR5 or CXCR4 receptors on the surface of T cell, and other susceptible cells in the body. If entry inhibitors are successful in blocking these proteins, HIV is unable to bind to the cell surface and gain entry. Fusion inhibitors, such as enfuvirtide (Fuzeon or T-20), are newer treatments that work by blocking HIV entry into cells. Fuzeon is a peptide that binds to gp41 preventing HIV from binding to the surface of T cells and was approved by the FDA in 2003. There are other experimental drugs that show early promise as well. PRO-542 and TNX-355 target the CD4 protein, and vicriviroc and maraviroc target the CCR5 protein used as coreceptors to enter cells.

Integrase Inhibitors

A critical step in the HIV life cycle is the integration of the virus's genetic information into the host cell DNA. This process enlists the host cell as a "HIV factory," which is programmed to produce numerous virions each hour. Integrase is the enzyme that accomplishes this task. Like proteases and reverse transcriptase, integrase also plays a vital role in the retroviral life cycle, which makes this enzyme an attractive target for the development of new anti-AIDS agents. There have been several attempts to create integrase inhibitors over the past 10 years. Successful integrase inhibitors are oligonucleotides, which are small segments of DNA or RNA that are synthetically prepared. Modified oligonucleotides can serve to block RNA/DNA interactions, thus modifying protein or enzyme synthesis. One drawback to integrase inhibitors is that they only have one chance to act for each cell. If an inhibitor fails, any further attempts are futile since the genetic information would already be incorporated. MK-0518 is an experimental drug from Merck & Co. that targets integrase. It is the first integrase inhibitor to go to trial. MK-0518 is taken orally twice daily and doses of 200, 400, and 600 mg are being studied. The drug is now entering phase three clinical trials.

HAART

With the new knowledge that the virus can mutate and that multiple drugs might need to be administered at the same time to get the desired effect, **highly active antiretroviral therapy** (HAART) became the recommended treatment for HIV infection in 1996. This is also called **combination therapy** (Figure 9-2. shows selected antiretrovials). HAART combines three or more anti-HIV medications in a daily regimen, sometimes referred to as a "cocktail," thereby reducing the chances of viral mutation against all of these drugs.

So far, the combination HAART treatment is the closest thing medical science has to an effective therapy. The key to its success in some patients lies in its ability to disrupt HIV at different stages in its replication cycle. Reverse transcriptase inhibitors, which usually make up two drugs in the HAART regimen, inhibit an enzyme crucial to the early stage of HIV replication (RNA to DNA). Protease inhibitors hold back another enzyme that functions near the end of the HIV replication process. HAART has been credited as a major factor in significantly reducing the number of deaths from AIDS, especially in the United States and other developed countries where access to treatment is not restricted by cost. While HAART is not a cure for AIDS, it has greatly improved the health of many people with AIDS and it reduces the amount of virus circulating in the blood to nearly undetectable levels. Researchers, however, have shown that HIV remains present in hiding places, such as the lymph nodes, brain, testes, and retina of the eye, even in people who have been treated.

Figure 9-2 A selection of HIV/AIDS antiretroviral drugs. (Image © Mircea Bezergheanu. Used under license from Shutterstock, Inc.)

Although HAART has several beneficial effects, there are several side effects associated with the use of these drugs and many of these effects may be severe. Some of the NRTIs may cause a decrease of red and/or white blood cells (pancytopenia), especially when taken in the later stages of the disease. Some may also cause inflammation of the pancreas and painful nerve damage. This reduction in red blood cells can lead to anemia, increased bruising or bleeding, and an inability to efficiently clot the blood. These latter effects stem from the fact that the blood platelets that are responsible for the initial stages of the clotting process are greatly reduced. There have been reports of complications and other severe reactions, including death, to some of the antiretroviral nucleoside analogs when used alone or in combination. Therefore, any patient taking these drugs should be closely monitored for signs of these side effects. Protease inhibitors also have associated side effects. The most common side effects associated with protease inhibitors include diarrhea and nausea. In addition, there is also a risk of drug interactions when taking protease inhibitors, which may result in serious side effects. Fuzeon inhibitors may also cause severe allergic reactions such as pneumonia, labored breathing, chills, fever, skin rash, blood in urine, vomiting, and low blood pressure. There have also been reports of local skin reactions at the injection site (it is given as an injection underneath the skin). Researchers and physicians alike have however determined that the benefits outweigh the risks associated with treatment using HAART and other anti-HIV drugs. AIDS killed an estimated 2.1 million people in 2007. If everyone had access to ARV therapy then the death toll would be much lower.

Atripla and AIDS Treatment

On July 12, 2006, the FDA approved Atripla tablets, a fixed-dose combination of three widely used antiretroviral drugs, in a single tablet taken once a day, alone or in combination with other antiretroviral products for the treatment of HIV-1 infection in adults. Atripla, the first one-pill, once-a-day product to treat HIV/AIDS, combines the active ingredients of Sustiva (efavirenz), Emtriva (emtricitabine), and Viread (tenofovir disoproxil fumarate). Each component of Atripla is currently approved for use in combination with other antiretroviral agents to treat HIV-1 infected adults. The safety and effectiveness of each component were previously demonstrated in clinical trials to support their individual approval. In addition, the safety and effectiveness of the combination of these three drugs were shown in a 48 week-long clinical study with 244 HIV-1 infected adults receiving the drugs contained in Atripla. The approval of Atripla simplifies the treatment regimen for HIV-1 infected adults, and will potentially improve the ability of patients to adhere to treatment resulting in long-term effective control of HIV-1. In the clinical trial phase, 80% of the participants achieved a marked reduction in HIV viral load and a substantial increase in the number of healthy CD4 cells needed to combat the infection. It is hoped that Atripla will significantly simplify the drug treatment regimen, helping to increase adherence, thus reducing potential development of viral resistance to the drugs. Like the individual drugs that make up this single dose therapy, Atripla is not without side effects. The most common side effects include headache, dizziness, abdominal pain, nausea, vomiting, and development of rash. Other potentially serious adverse events reported for the use of Atripla's ingredients include liver toxicity, renal impairment, and severe depression in a subset of patients (*http://www. fda.gov/cder/drug/info page/atripla/factsheet.htm*).

New drug therapies and new trials are being developed every day in the labs of various researchers, some of whom have dedicated their entire careers to HIV research. Those who have experienced the rapid developments in this field over the last two decades will admit that it has been quite a roller-coaster ride. The dream of a cure or at least eradication of the virus from the body of an infected person that was widely hoped for in the beginning eventually had to be abandoned. In 1997, it was estimated that viral suppression using antiviral drugs for a duration of three years was necessary to completely eradicate the virus from an individual. Researchers arrived at this conclusion by a combination of mathematical models and lab observations. The thought was that after this period, all infected cells would presumably have died. Since then, the duration of three years has consistently been adjusted upward and newer studies came to the sobering conclusion that HIV remains detectable in latent infected cells, even after long-term suppression.

To date, nobody knows how long these latent infected cells survive, and all indications are that even a small number of them would be sufficient for the infection to flare up again as soon as treatment is interrupted. The most recent estimates for eradication of HIV-infected cells from the body are approximately 50–70 years. One thing is certain, HIV will not be curable for at least the next ten years. Dr. Michael Gottlieb described it best in a *Los Angeles Times* article when he said, "I've always looked at AIDS therapy as a series of leaky lifeboats. You stay in the first one until you're sinking, then jump to another one. But you don't give up looking for others." HIV/AIDS has not been a respecter of anyone. It has crossed racial, gender, socioeconomic, religious, regional, and continental lines. While it is true that the disease has been devastating in areas such as sub-Saharan Africa, it is also very prevalent in MSM and injection-drug users in the United States, and literally in all of our backyards. Instead of eradication, it has become more realistic to consider the lifelong management of HIV infection as a chronic disease in a similar manner to diabetes mellitus. This will, however, mean that patients will need to have the discipline and ability, both mentally and physically, to take the currently available pills several times daily at fixed times for the next 10, 20, or even 30 years. Regardless, HIV remains a dangerous and cunning opponent on which we cannot afford to turn our backs. The developments in the area of treatment and management of the disease over the next five to ten years will be very important in shaping the future of this deadly epidemic.

CHAPTER 10

HIV Vaccine: A Futile Search?

Even with the great success and increasing availability of antiretroviral drugs, which have been shown to significantly extend and improve the quality of life of HIV-infected persons, these drugs do not "cure" the HIV-infected patient. Physicians and their patients alike would be much happier about a drug that is capable of completely preventing HIV infections. While condom use, abstinence, and needle exchange programs have been very effective in significantly reducing the rate of HIV infection in many areas over the past decade, these practices are totally dependent on responsible cooperation of all sexually active individuals. As we all know, human behavior is not easily controlled and people will do whatever they want despite the possible consequences. A strategy that would completely prevent HIV infection is therefore highly desirable.

Almost every time I talk with students, family members, and friends about HIV infection, viral mutation, drug resistance, the prevalence of the disease, and the implications for the next generation, I get asked the all-important question, "How far are we away from making an HIV vaccine?" The fact is that there is currently no vaccine to prevent HIV. Researchers, however, continue to develop and test potential HIV vaccine targets with the hope that they will in the near future develop a vaccine that can protect people from HIV infection, or at least lessen the chance of getting HIV or AIDS should a person be exposed to the virus.

How Do Vaccines Work?

What is a vaccine anyway and why is it such a desirable treatment option for HIV/AIDS? A vaccine is a medical product that is designed to stimulate the body's immune system in order to prevent or control an infection. An effective preventive vaccine primes the immune system to fight off a particular microorganism so that it can't establish a serious infection or make you sick. A vaccine preparation therefore contains an antigen that may consist of whole disease-causing organisms (killed or weakened) or parts of such organisms, which is used to confer immunity against the disease that the organisms cause. Vaccine preparations can be natural, synthetic, or derived by recombinant DNA technology. These vaccines are typically delivered orally, intramuscularly (See Figure 10-1), nasally, or transcutaneously. There are three main ways to make a vaccine, although most strategies for an HIV/AIDS vaccine focus on a single strategy.

One can make what is known as a **killed virus vaccine.** This would be prepared by taking HIV particles and "killing" them with chemical, UV light, or heat treatments. This killed product is then mixed with an agent that helps to boost the immune system. While this strategy has been successful in the past in creating vaccines like the Jonas Salk polio vaccine, so far it has not been successful in AIDS vaccine development. A second strategy would be to make a **live attenuated vaccine.** In this case the viruses would still be alive, but would be weakened in order to reduce their ability to cause disease, but just enough to still allow it to induce immune responses inside the vaccine recipient. As you might imagine, with this approach there is the danger of getting HIV infection from the vaccine and, consequently, this approach has not been explored. As Dr. Emilio Emini, head of vaccine development at

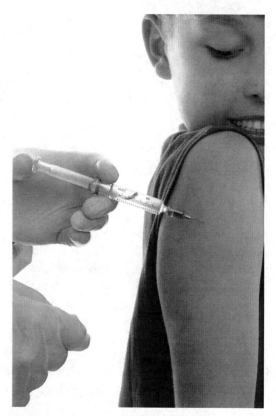

Figure 10-1 An effective preventative vaccine against HIV is the great hope for this century. (Image © Leah-Anne Thompson. Used under license from Shutterstock, Inc.)

Wyeth, explains, "In the case of live attenuated HIV, that virus never goes away because it becomes an integral part of the genetic information of the cells in that individual. Eventually it will cause disease." This does not mean that researchers have run out of options quite yet.

The third approach to vaccine development is the most recent and the safest of the three. It employs a **genetically engineered** or **subunit vaccine.** Subunit vaccines do not involve the use of the whole virus particles, but are constructed with two or more viral proteins. With this approach, vaccinologists can manipulate the nucleic acid (genetic material) of the pathogen in several ways. Certain integral genes whose products stimulate an immune response may be inserted into some sort of vehicle, typically another virus that does not cause disease in humans. This genetically engineered virus is injected into the person where it triggers an immune system. This live vector vaccine is able to stimulate a much stronger immune response than killed viruses. Alternately, they could create a vaccine by injecting only certain elements of the virus, such as portions of its glycoproteins, or synthesized pieces of viral protein. The bottom line in this approach is that the immune system would recognize and respond to these viral components, building up memory B and T cells ready to respond in the event that a real HIV particle is introduced to the body. The downside to this approach is that the immune system does not get to sample the entire virus, but is limited to the components in the vaccine mixture. This approach is, however, very safe. Most current HIV/AIDS vaccine strategies employ genetically engineered vaccine strategies.

The Challenges of Creating an HIV Vaccine

It is no secret that vaccine development has always been a long and arduous process. However, several factors make HIV a particularly challenging virus to create a vaccine against. The main problem that HIV vaccine developers face is the unusual nature of HIV. The virus not only evades the body's immune system response, it actually destroys the very immune system cells (the CD4 T cells) that are central to the body's defense against a viral or bacterial infection. As soon as the virus gains entry into the CD4 cells, it inserts its genetic material into the cell's genome and literally hijacks the cells protein-making machinery to create more viruses. If a vaccine is able to stimulate the immune response and the virus is not being killed, this could actually spur further virus production. To compound an already bad situation, HIV, like all other retroviruses, must convert its genetic code from RNA to DNA once inside the CD4 cell. It turns out that this process is notoriously error-prone. These translational errors often lead to high viral mutation rates. Since the immune system typically defends the body by producing highly specific antibodies (See Figure 4-2), mutated viruses would not be recognized and would therefore be able to escape the immune response stimulated by a genetically engineered subunit type vaccine. The vaccine-resistant virus strains could therefore be transmitted to others. Within HIV-1 alone, there are at least 10 different subtypes (as well as recombinant viruses combining several subtypes), which are found in different parts of the globe (See Figure 3-1). Researchers must therefore develop multiple vaccines or one that works optimally across multiple subtypes.

Even if researchers could solve all of the problems and work out all of the issues cited above, they would still have to face the greatest challenge yet in creating an efficacious AIDS vaccine; *that is, in the past 27 years since the AIDS pandemic started, doctors have not found a single patient who was able to rid himself/herself of the virus*. This is an extremely discouraging problem facing vaccine researchers today. To date, even with the deadliest of diseases caused by viruses such as smallpox, polio, measles, or bacterial infections caused by *Salmonella* species, *Mycobacterium tuberculosis*, and *Escherichia coli* (with only a few exceptions—herpes), when one gets infected, there's a certain level of mortality; then you clear the infection, the body mounts an immune response, and it's unlikely you're going to get infected again. Once you get infected with HIV, the body seems to be completely incapable of eliminating the virus. Over the years, successful vaccine development has always taken its queue from what the body did to protect itself under natural circumstances. Researchers then develop a vaccine that mimics what the body has done in order to provide some initial help to the system so it does not get overwhelmed. With HIV the body has not been successful in mounting a response that clears the virus from the system; therefore, there is no real framework on which to base vaccine development. Despite these nearly insurmountable challenges, the discovery and development of safe, efficacious, and cost-effective vaccines to prevent HIV infection and/or disease has become the mission of several research groups worldwide (*http://www.iavi.org/*).

Preventative vs. Therapeutic Vaccines

When most people think about a vaccine against HIV, what they are envisioning is a product that will prevent infection in the first place since the virus cannot be cleared. The fact is that there is no vaccine available for any disease that prevents the organism from entering the body or cells in the first place. Vaccines typically prevent disease and not infection.

A **preventative vaccine** is designed to block infection in people who are HIV negative. When our bodies encounter a microbe of any type, the immune system responds with all its arsenals with two important goals—first, to eliminate the threat, and second, to build an immunologic memory so that it continues to "remember" how the microbe looked and can quickly defeat the invader should it ever try to infect again. Since the vaccine is designed to look like the real microorganism, it trains the immune system to recognize and attack the real microorganism should it ever be encountered by the vaccinated individual. Therefore, if you have received an effective protective vaccine, your immune system will "remember" how to quickly attack and defeat that particular microorganism for many years. The bulk of early HIV vaccine research was focused on the development of a preventative vaccine, which tried to encourage the immune system to produce **antibodies** that would block the virus from infecting cells. This approach was quite successful in laboratory experiments; however, they all failed when mutated strains of HIV were introduced to the culture. According to the **AIDS Vaccine Advocacy Coalition** (AVAC), determining how to stimulate an antibody response is "a task that most researchers consider essential for an optimal vaccine but one that's proven impossible so far" (*http://www.avac.org/*).

The current research efforts to develop a vaccine to combat HIV have shifted to **therapeutic vaccines,** which seem to have a greater potential for success. Most of the 30 vaccine candidates that are currently in clinical trials are therapeutic vaccines. A therapeutic HIV vaccine (also known as a treatment vaccine) is a vaccine used in the treatment of an HIV-infected person. Therapeutic HIV vaccines are designed to boost the body's immune response to HIV in order to better control the infection. This type of vaccine is not designed to prevent HIV infection; however, it is hoped that they will significantly delay the progression to AIDS, thereby greatly improving the quality of life for HIV-infected persons. There are currently no FDA-approved therapeutic vaccines; however, there are several prospective ones in clinical trial to determine if they are safe and effective in treating HIV infections. It is hoped that if therapeutic vaccines are able to strengthen the body's natural anti-HIV immune response, people with HIV will not have to rely so heavily on the antiretroviral drugs currently used to treat HIV infection. Currently, HIV patients must continue to take antiretroviral drugs for life, and most of these drugs cause serious and sometimes life-threatening side effects. It should be noted that even an effective therapeutic HIV vaccine probably will not be able to replace antiretroviral drugs entirely. At best, a therapeutic HIV vaccine may help control HIV infection and keep people healthy while minimizing the need for antiretroviral drugs.

There is no doubt that tremendous progress has been made in combating HIV infection in the United States and other industrialized countries. However, new infections continue to occur worldwide. There is, however, an urgent need to find effective measures to prevent new infections, particularly in those parts of the world where HIV continues to spread unabated, and antiretroviral therapies are outside of the financial reach of those in greatest need. Throughout history, vaccines have always been great equalizers in society because they were always made available to both the rich and the very poor—those in developed regions and those in the poverty stricken corners of our planet. While a significant portion of HIV research dollars has been appropriated for vaccine research, most researchers agree that a successful vaccine is no closer than decades away. "Even if we come up with a cure or vaccine tomorrow, just think about the time that would be needed to implement all these measures widely throughout the world," adds Dr. David Ho, director and CEO of the Aaron Diamond AIDS Research Center. "And even with that optimal scenario, it would be decades before this fight is won, and we're certainly, unfortunately, not in that situation" (*http://www.pbs.org/wgbh/ pages/frontline/aids/virus/vaccines.html*).

APPENDIX A

Current Selected Issues and Future Outlook

HIV/AIDS and Drug Abuse

After 27 years of devastation in the human population, HIV/AIDS remains a potentially deadly chronic disease. Ninety percent of HIV-positive people worldwide don't know that they have HIV. The disease continues to wreak havoc in African American communities in the United States mainly because of denial and a lack of education about HIV/AIDS (Figure A-1). More frightening is mounting evidence suggesting that the disease is making a comeback among the group that was most educated about AIDS even in its early days—the gay community. The number of new infections nationwide among gay men rose every year between 2000 and 2006. This dramatic increase seems to be directly linked to the use of crystal methamphetamine. Crystal meth use has been shown to double the risk of HIV infection in any single sexual encounter. This may be because users are more likely to be unsafe; however, research is also showing that the body of a person high on meth suffers physiological changes that make transmission easier.

Prevention is currently the only cure for HIV. Effective prevention measures will, however, require continued commitment from every person at risk, as well as those who are currently infected, their family members, and society as a whole. These prevention efforts need to keep pace with the changing face of the HIV epidemic. An important facet of this prevention strategy will have to include the younger members of our society, who might not remember or were not born at the time when HIV/AIDS was very deadly, and there was uncertainty and chaos. The younger generation needs to receive basic HIV-prevention messages on a continuous basis. It is clear that much work remains to be done on the search for an efficacious vaccine, whether it is therapeutic or preventative. The ideal situation would entail the development of an effective preventative vaccine at a cost that is affordable for all who need it. This, however, might just be wishful thinking at this stage.

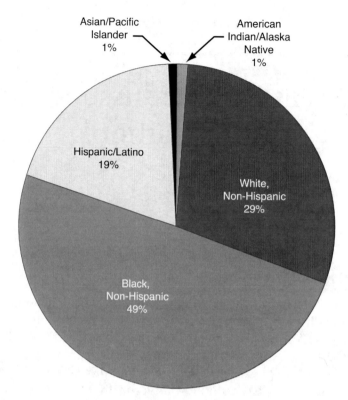

Figure A-1 Race/ethnicity of persons (including children) with HIV/AIDS diagnosed during 2007. (Source: CDC fact sheet, *http://www.cdc.gov/HIV/topics/ aa/resources/ factsheets/aa.htm*)

Racial and Ethnic Disparities in Infection Rate

Racial and ethnic minorities have been disproportionately affected by HIV/AIDS since the beginning of the epidemic, and minority Americans now represent the majority of new AIDS cases (71%) and of those estimated to be living with AIDS (64%) in the United States in 2007. African Americans and Latinos continue to account for a disproportionate share of new AIDS diagnoses (Figure A-1). The 2007 U.S. statistics indicate that African Americans have the highest AIDS case rates of any racial/ethnic group, followed by Latinos, American Indian/Alaska Natives, Whites, and Asian/Pacific Islanders. The AIDS case rate per 100,000 of the population for African Americans was over 9 times that of Whites in 2006. Although African Americans make up only 13% of the U.S. population, they accounted for 49% of all AIDS cases in 2007.

In response to this growing trend, the CDC has established the **African American Working Group** to focus on the urgent issue of HIV/AIDS in African Americans. The mandate of the working group is to develop a comprehensive response that will guide CDC's attempt to strengthen HIV/AIDS prevention and intervention activities directed toward African Americans. The CDC is already engaged in a wide range of activities that involve leaders in the African American community who will work together to decrease the incidence of HIV/AIDS. Also at work is the CDC's **Advancing HIV Prevention program** (AHP) (*http://www.cdc.gov/HIV/topics/prevprog/AHP/default.htm*), an initiative that aims to reduce barriers to early diagnosis of HIV infection while increasing access to quality medical care, treatment, and prevention services for those living with HIV. AHP proposes to make HIV testing a routine part of medical care, implement new models for diagnosing HIV infections outside of clinical settings, prevent new infections by working with HIV-infected persons and their partners, and continue with efforts to decrease perinatal HIV transmission.

Routine HIV Testing as Part of Regular Medical Care

The CDC estimates that between 25% and 30% of the roughly 1 million HIV-infected Americans are unaware of their status and therefore are unknowingly transmitting the virus. Current studies seem to indicate that HIV-infected persons access the health care system but are not tested for HIV until late in the course of their disease. For many, it is too late to benefit from the effective HIV treatments that are available today. The CDC is therefore expected to recommend that every American between ages 13 and 64 be tested for HIV at least once during their lifetime. The CDC will most certainly go further in recommending that testing should be made a standard part of medical care regardless of an individual's risk factors or HIV prevalence in the community. These recommendations, which are currently in draft form, are a result of extensive consultation with representatives from professional organizations, state and federal agencies, community-based organizations, people living with HIV/AIDS, health departments, and academia.

Knowledge of one's status allows for earlier access to effective treatment that results in a longer, healthier life span. When people learn that they are HIV infected, the necessary steps can then be taken to protect their partners. In the past, knowledge of positive HIV status has been shown to reduce risky behavior with uninfected partners by approximately 68%. CDC estimates that most of the new HIV infections in the United States are transmitted by the 25% of people who do not even realize that they are infected. The hope is that by expanding access to HIV testing and making it a standard part of medical care, those who are not aware that they are infected may be effectively reached, and will in turn take steps to protect their partners. Some argue, however, that while broader testing is necessary, it is not enough to stem the tide of the HIV epidemic in our population.

APPENDIX B

Selected HIV/AIDS Online Resources

http://www.unfpa.org/aids_clock/index.html#: **The AIDS Clock** was created in 1997 and became web-based in 2000. The United Nations Population Fund created the clock as a way to acknowledge both the toll of the epidemic, and the partnership that was formed to tackle it, **UNAIDS** (the United Nations Joint Program on HIV/AIDS). While the precise numbers of people living with HIV, people who have been newly infected, or who have died of AIDS are not known, the **AIDS clock** provides point estimates in real time along with other useful statistics on AIDS.

http://www.aidsinfo.nih.gov/: **AIDS***info* is a U.S. Department of Health and Human Services (DHHS) website that offers the latest federally-approved information on HIV/AIDS clinical research, treatment and prevention, and medical practice guidelines for people living with HIV/AIDS, their families and friends, health care providers, scientists, and researchers.

http://www.thebody.com/anin/aninpage.html: **AIDS National Interfaith Network (ANIN)** is a private, non-profit organization founded in 1988 to ensure that individuals with HIV and AIDS receive compassionate and nonjudgmental support, care, and assistance. ANIN coordinates a network of nearly 2,000 AIDS ministries. ANIN works with national faith-based, AIDS-specific networks; supporting community-based AIDS ministries; and educates AIDS service organizations, the religious community at large, and the general public about AIDS ministries. The site is hosted by *The Body.*

http://www.amfar.org/cgi-bin/iowa/index.html: **The American Foundation for AIDS Research (amfAR)** is the nation's leading nonprofit organization dedicated to the support of HIV/AIDS research. The organization seeks to develop prevention methods (including a vaccine), improve treatments, and ultimately find a cure for AIDS. amfAR-funded research makes significant contributions to the lives of people with HIV/AIDS and to the global effort to arrest the epidemic.

http://www.avert.org/: **AVERT**, an international HIV and AIDS charity based in the United Kingdom, has a number of overseas projects, helping with the problem of HIV/AIDS in countries where there is a particularly high rate of infection, such as

South Africa, or where there is a rapidly increasing rate of infection, such as in India. This site is very comprehensive with up-to-date statistics and touches on all of the important issues.

http://www.hivinsite.org/: **HIV InSite** is a source for comprehensive, in-depth HIV/AIDS information and knowledge developed by the Center for HIV Information (CHI) at the University of California San Francisco (UCSF), *http://chi.ucsf.edu/*.

http://www.nytimes.com/library/national/science/aids/aids-index.html: **The** *New York Times* hosts this wonderful site containing current relevant news about HIV/AIDS as well as audio and video multimedia resources organized by dates in various categories. There is also a great archive of articles published by the *New York Times* over the years.

http://www.thebody.com/index.html: **The Body** is a comprehensive site with research-based HIV/AIDS resources, including treatment and prevention, as well as forums where you can ask relevant questions to experts in the field. The Body also hosts the "Visual AIDS Web Gallery" with art created by HIV-positive artists.

http://www.cdc.gov/hiv/dhap.htm: **The Center for Disease Control and Prevention (CDC)** is the major United States government-run resource for facts about HIV/AIDS in the United States. The site includes information on prevention programs, fact sheets about different high-risk groups, and statistics broken down by state. This is the site on which the first cases of HIV infection were published and it continues to host a collection of articles from the CDC's Morbidity and Mortality Weekly Report on HIV/AIDS.

http://www.iavi.org/: **The International AIDS Vaccine Initiative (IAVI)** is a global not-for-profit organization working to speed the search for a vaccine to prevent HIV infection and AIDS. Founded in 1996 and operational in 23 countries, IAVI and its network of partners research and develop vaccine candidates. IAVI also advocates for a vaccine to be a global priority and works to assure that a future vaccine will be accessible to all who need it.

http://www.unaids.org/: **UNAIDS, the Joint United Nations Program on HIV/AIDS,** brings together the efforts and resources of 10 UN system organizations to the global AIDS response. The site brings to the forefront the many challenges facing individuals with AIDS as well as the progress being made in education, treatment, and new initiatives.

http://www.worldaidscampaign.info/: **The World AIDS Campaign (WAC)** is an orga-nization that has been established to strengthen and connect together the advocacy and campaign activities targeting governments and other AIDS organizations. The vision of the WAC is to establish a global movement bringing renewed impetus and resolve to the fight against the epidemic. The WAC is currently responsible for organizing the annual **World AIDS Day** on December 1st each year.

ENDNOTES

1. CDC, C.f.D.C.a.P.-. *Pneumocystis Pneumonia—Los Angeles. MMWR*, 1981. **30**(21): pp. 1–3.

2. Littman, D. R., *Chemokine receptors: keys to AIDS pathogenesis?* Cell, 1998. **93**(5): pp. 677–80.

3. Duesberg, P., *HIV is not the cause of AIDS*. Science, 1988. **241**(4865): pp. 514, 517.

4. Duesberg, P., and D. Rasnick, *The AIDS dilemma: Drug diseases blamed on a passenger virus*. Genetica, 1998. **104**(2): pp. 85–132.

5. Delaney, M., *HIV, AIDS, and the Distortion of Science, in Focus 2000*. 2000, Project Inform, San Francisco, California.

6. Lederer, B., *Dead Certain?* April 2006, POZ Magazine.

7. Mellors, J. W., et al., *Plasma viral load and CD4+ lymphocytes as prognostic markers of HIV-1 infection*. Ann Intern Med, 1997. **126**(12): pp. 946–54.

8. Ciesielski, C. A., et al., *The 1990 Florida dental investigation. The press and the science*. Ann Intern Med, 1994. **121**(11): pp. 886–8.

9. O'Brien, S. J., and J. J. Goedert, *HIV causes AIDS: Koch's postulates fulfilled*. Curr Opin Immunol, 1996. **8**(5): pp. 613–8.

10. CDC, C.f.D.C.a.P.-. *HIV AIDS Surveillance Report*. 1999. **Year-end edition Vol. 11, No. 2:** pp. 25–27.

11. Gao, F., et al., *Origin of HIV-1 in the chimpanzee Pan troglodytes troglodytes*. Nature, 1999. **397**(6718): pp. 436–41.

12. CDC, C.f.D.C.a.P.-. *HIV/AIDS Surveillance Report*. 2004. **15:** pp. 1–46.

13. Halkitis, P. N., et al., *Barebacking identity among HIV-positive gay and bisexual men: Demographic, psychological, and behavioral correlates*. Aids, 2005. **19 Suppl 1:** pp. S27–35.